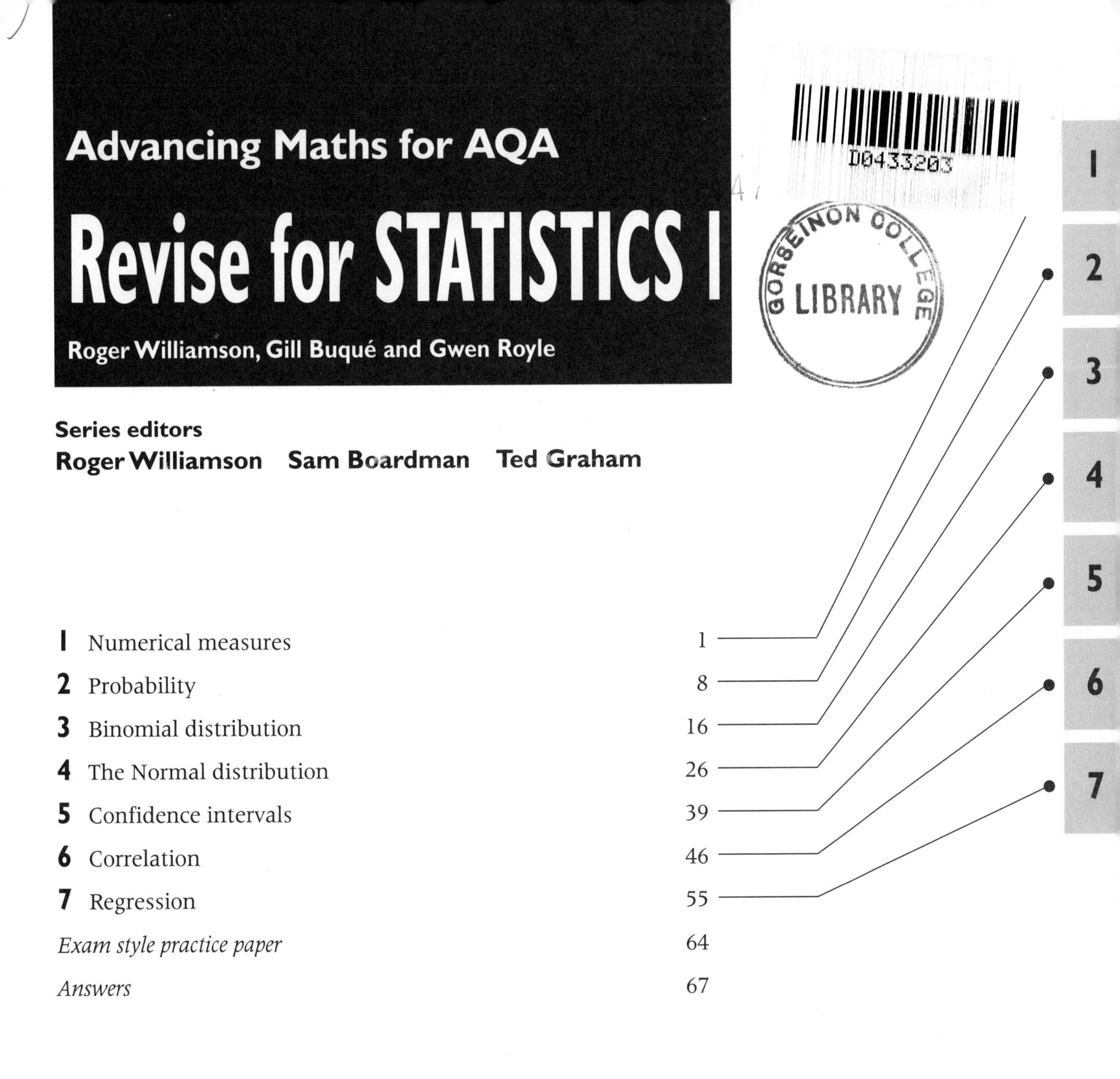

Advancing Maths for AQA

Revise for STATISTICS 1

Roger Williamson, Gill Buqué and Gwen Royle

Series editors
Roger Williamson Sam Boardman Ted Graham

Heinemann Educational Publishers
Halley Court, Jordan Hill, Oxford OX2 8EJ
Part of Harcourt Education

Heinemann is the registered trademark of
Harcourt Education Limited

First published 2005

08 07 06 05
10 9 8 7 6 5 4 3 2 1

British Library Cataloguing in Publication Data is available from the British Library on request.

10-digit ISBN: 0 435513 55 9
13-digit ISBN: 978 0 435513 55 9

Typeset and illustrated by Tech-Set Limited, Gateshead, Tyne & Wear
Original illustrations © Harcourt Education Limited, 2005
Cover design by mccdesign ltd
Printed in the United Kingdom by Scotprint

Acknowledgements
The publishers' and authors' thanks are due to AQA for permission to reproduce questions from past examination papers.

The answers have been provided by the authors and are not the responsibility of the examining board.

Every effort has been made to contact copyright holders of material reproduced in this book. Any omissions will be rectified in subsequent printings if notice is given to the publishers.

About this book

This book is designed to help you get your best possible grade in your Statistics 1 examination. The authors are Chief and Principal examiners, and have a good understanding of AQA's requirements.

Revise for Statistics 1 covers the key topics that are tested in the Statistics 1 exam paper. You can use this book to help you revise at the end of your course, or you can use it throughout your course alongside the course textbook, *Advancing Maths for AQA AS & A-level Statistics 1*, which provides complete coverage of the syllabus.

Helping you prepare for your exam

To help you prepare, each topic offers you:

- **Key points to remember** – summarise the statistical ideas you need to know and be able to use.
- **Worked examples** – help you understand and remember important methods, and show you how to set out your answers clearly.
- **Revision exercises** – help you practise using these important methods to solve problems. Exam-level questions are included so you can be sure you are reaching the right standard, and answers are given at the back of the book so you can assess your progress.
- **Test Yourself questions** – help you see where you need extra revision and practice. If you do need extra help, they show you where to look in the *Advancing Maths for AQA AS & A-level Statistics 1* textbook and which example to refer to in this book.

Exam practice and advice on revising

Examination style paper – this paper at the end of the book provides a set of questions of examination standard. It gives you an opportunity to practise taking a complete exam before you meet the real thing. The answers are given at the back of the book.

How to revise – for advice on revising before the exam, read the How to revise section on the next page.

How to revise using this book

Making the best use of your revision time

The topics in this book have been arranged in a logical sequence so you can work your way through them from beginning to end. But **how** you work on them depends on how much time there is between now and your examination.

If you have plenty of time before the exam then you can **work through each topic in turn**, covering the key points and worked examples before doing the revision exercises and test yourself questions.

If you are short of time then you can **work through the Test Yourself sections** first, to help you see which topics you need to do further work on.

However much time you have to revise, make sure you break your revision into short blocks of about 40 minutes, separated by five- or ten-minute breaks. Nobody can study effectively for hours without a break.

Using the Test Yourself sections

Each Test Yourself section provides a set of key questions. Try each question:

- If you can do it and get the correct answer, then move on to the next topic. Come back to this topic later to consolidate your knowledge and understanding by working through the key points, worked examples and revision exercises.
- If you cannot do the question, or get an incorrect answer or part answer, then work through the key points, worked examples and revision exercises before trying the Test Yourself questions again. If you need more help, the cross-references beside each Test Yourself question show you where to find relevant information in the *Advancing Maths for AQA AS & A-level Statistics 1* textbook and which example in *Revise for S1* to refer to.

Reviewing the key points

Most of the key points are straightforward ideas that you can learn: try to understand each one. Imagine explaining each idea to a friend in your own words, and say it out loud as you do so. This is a better way of making the ideas stick than just reading them silently from the page.

As you work through the book, remember to go back over key points from earlier topics at least once a week. This will help you to remember them in the exam.

CHAPTER 1

Numerical measures

Key points to remember

1 The three most common measures of 'average' are the **mean, median** and **mode**.

2 The most commonly used is the mean. A **sample mean** is denoted $\bar{x}$ and a **population mean** is denoted μ.

3 The mean is calculated using the formula $\frac{\Sigma x}{n}$.

4 The three most common measures of spread are the **range, interquartile range** and **standard deviation**.

5 The most commonly used measure of spread is the standard deviation. A **population standard deviation** is denoted σ.

$$\sigma = \sqrt{\frac{\Sigma(x - \mu)^2}{n}}$$

6 The data you deal with will usually be a sample. If this is so it is not possible to calculate σ. You should estimate σ by s.

$$s = \sqrt{\frac{\Sigma(x - \bar{x})^2}{n - 1}}$$

7 You should practise obtaining the mean and standard deviation directly using your calculator. This saves a lot of time and is acceptable in the examination.

8 The standard deviation squared is called the variance.

9 If a variable is increased by a constant amount its average will be increased by this amount but its spread will be unchanged. This applies whichever measures of average and spread are used.

10 If a variable is multiplied by a constant amount both its average and spread will be multiplied by this amount. This is true whichever measures of average and spread are used.

Worked example 1

Andrea keeps free-range chickens and collects their eggs every morning. The numbers of eggs she collects on 12 mornings are:

13 7 12 14 10 8 12 7 12 6 11 14

(a) Find the mode, median and mean of the daily number of eggs collected.

(b) Find the range and the interquartile range of the daily number of eggs collected.

(c) Find the standard deviation of the daily number of eggs collected.

(a) To find the mode and the median, arrange the numbers in order of size.

6 7 7 8 | 10 11 12 12 | 12 13 14 14

The most frequently occurring number is 12 so mode = 12 eggs per day.

There are two middle values so take the average giving median $= \frac{11 + 12}{2} = 11.5$ eggs per day.

The mean can be found by $\frac{\Sigma x}{n} = \bar{x} = \frac{126}{12} = 10.5$, giving mean = 10.5 eggs per day.

Using **3**

Use the statistical keys on your calculator to save time.

(b) Subtracting the smallest value from the largest gives range = 14 − 6 = 8 eggs per day.

The quartiles are $\frac{7 + 8}{2} = 7.5$ and $\frac{12 + 13}{2} = 12.5$ so interquartile range = 12.5 − 7.5 = 5 eggs per day.

(c) To answer part **(c)** using the formula we need to find $\Sigma(x - \bar{x})^2 = \Sigma(x - 10.5)^2$.

The values of $x - 10.5$ are

2.5 −3.5 1.5 3.5 −0.5 −2.5 1.5 −3.5 1.5 −4.5 0.5 3.5

and $\Sigma(x - 10.5)^2 = 89$.

$\sqrt{\frac{\Sigma(x - 10.5)^2}{12 - 1}} = \sqrt{\frac{89}{11}}$ giving standard deviation = 2.84 eggs per day to 3 s.f.

Using **6**

Press s or σ_{n-1} on your calculator.

$\sigma = \sqrt{\frac{89}{12}} = 2.72$ would also be accepted in the examination.

Worked example 2

Fifty mathematics students take a test and are each awarded a mark out of 20. Their marks are summarised in the following table.

Mark	7	8	9	10	11	12	13
Number of students	3	8	13	11	7	6	2

(a) Calculate:

(i) the mean, giving your answer to the nearest whole number,

(ii) the standard deviation, giving your answer to 1 d.p.,

(iii) the variance.

(b) The mathematics lecturer decides that the test was too hard and the marks should be increased. Calculate, to the same levels of accuracy as in part **(a)**, the mean, standard deviation and variance of the increased marks if:

(i) every student's mark is increased by 5 marks,

(ii) every student's mark is increased by 50%.

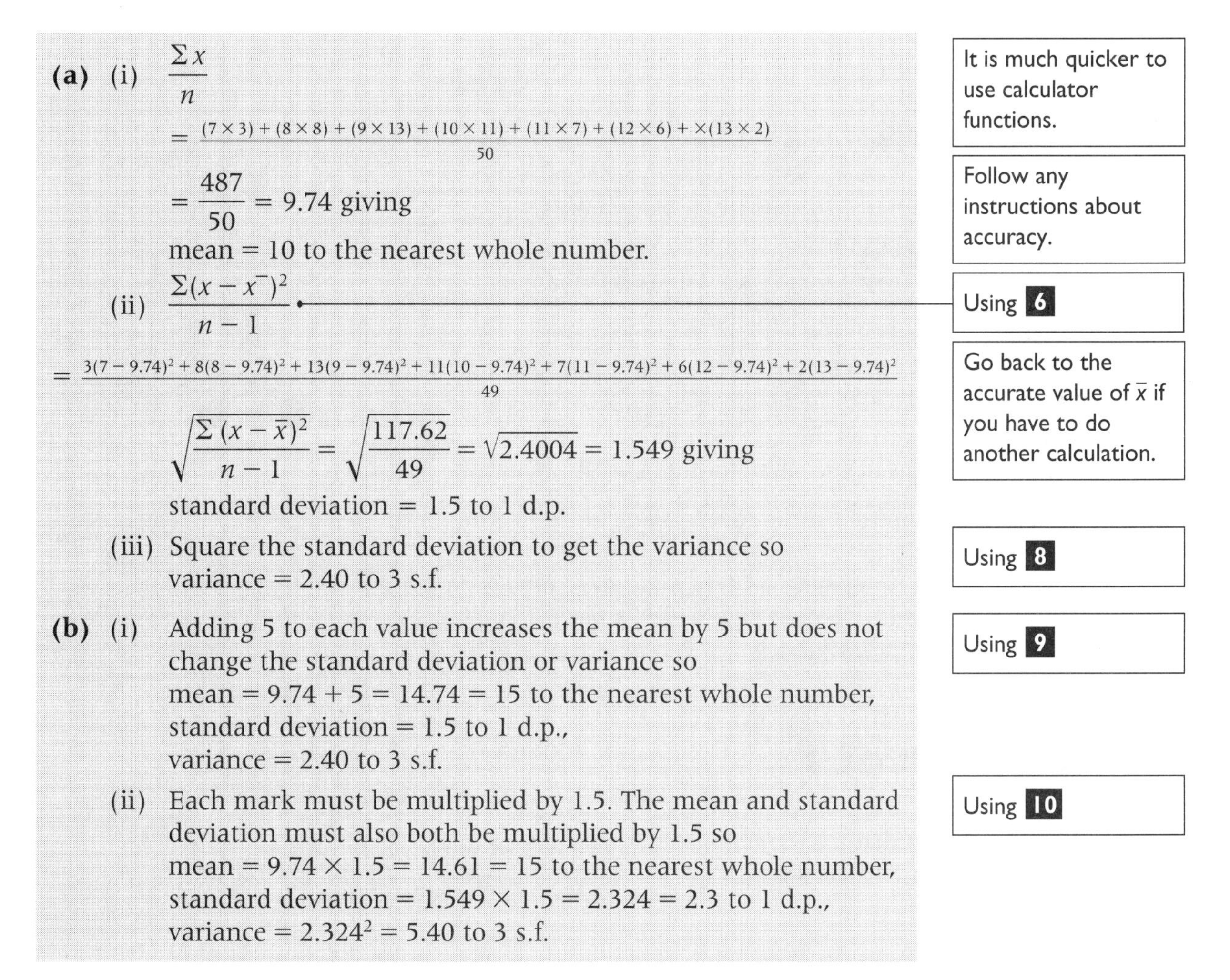

(a) (i) $\dfrac{\Sigma x}{n}$

$= \dfrac{(7\times3)+(8\times8)+(9\times13)+(10\times11)+(11\times7)+(12\times6)+\times(13\times2)}{50}$

$= \dfrac{487}{50} = 9.74$ giving

mean = 10 to the nearest whole number.

It is much quicker to use calculator functions.

Follow any instructions about accuracy.

(ii) $\dfrac{\Sigma(x-\bar{x})^2}{n-1}$

Using **6**

$= \dfrac{3(7-9.74)^2+8(8-9.74)^2+13(9-9.74)^2+11(10-9.74)^2+7(11-9.74)^2+6(12-9.74)^2+2(13-9.74)^2}{49}$

$\sqrt{\dfrac{\Sigma(x-\bar{x})^2}{n-1}} = \sqrt{\dfrac{117.62}{49}} = \sqrt{2.4004} = 1.549$ giving

standard deviation = 1.5 to 1 d.p.

Go back to the accurate value of $\bar{x}$ if you have to do another calculation.

(iii) Square the standard deviation to get the variance so variance = 2.40 to 3 s.f.

Using **8**

(b) (i) Adding 5 to each value increases the mean by 5 but does not change the standard deviation or variance so
mean = 9.74 + 5 = 14.74 = 15 to the nearest whole number,
standard deviation = 1.5 to 1 d.p.,
variance = 2.40 to 3 s.f.

Using **9**

(ii) Each mark must be multiplied by 1.5. The mean and standard deviation must also both be multiplied by 1.5 so
mean = 9.74 × 1.5 = 14.61 = 15 to the nearest whole number,
standard deviation = 1.549 × 1.5 = 2.324 = 2.3 to 1 d.p.,
variance = 2.324^2 = 5.40 to 3 s.f.

Using **10**

Worked example 3

Jake belongs to a junior athletics club and competes in the 800 metres. His times for a random sample of 24 practice runs are given in the following table.

Time, t seconds	**Number of runs**
120–	2
125–	3
127–	4
129–	6
131–	5
133–	1
135–145	3

(a) Calculate estimates of the mean and standard deviation of Jake's times for practice runs in the 800 metres, giving your answers to the nearest second.

(b) Vinod also competes in the 800 metres. His times in practice runs are estimated to have mean 128 seconds and standard deviation 7 seconds.
Compare Jake's and Vinod's performances in the 800 metres.

(a) We can only find approximate values for the mean and standard deviation as we do not have exact values for Jake's running times. First construct a frequency table using the mid-values of the class intervals.

Time, t seconds	122.5	126	128	130	132	134	140
Frequency	2	3	4	6	5	1	3

By calculator $\bar{t} = 130.375$, $s = 4.705$, giving estimated values of
mean = 130 seconds to the nearest second,
standard deviation = 5 seconds to the nearest second.

Using **3** and **6**

(b) Vinod's mean running time is lower than Jake's so we can deduce that Vinod is faster than Jake on average. However Vinod's times have a higher standard deviation than Jake's so we can deduce that Vinod's performance is more variable than Jake's.

REVISION EXERCISE 1

1 The total number of goals scored per match in a county hockey league is recorded for the twenty matches played in a season. The results are listed below.

1 4 3 0 6 3 2 1 5 3
0 9 4 4 2 6 3 4 1 3

Find the mode, median and mean of the number of goals scored per match.

2 (a) Find the range, interquartile range and standard deviation of the number of goals scored per match for the results given in question **1**.

(b) The score recorded as 9 was a mistake and should have been 7.

(i) Name the measure of spread that will decrease by 2 when the mistake is corrected. Give a reason for your answer.

(ii) Name the measure of spread that will stay the same when the mistake is corrected. Give a reason for your answer.

3 The Millhill housing estate consists of 52 houses. The number of people living in each house is recorded with the following results.

Number of people	1	2	3	4	5	6	7	8
Number of houses	2	6	8	14	10	8	3	1

Use your calculator to find the mean and standard deviation of the number of people per house on the Millhill estate.

4 There are 76 children under the age of 15 living on the Millhill estate. Their ages are shown in the table below.

Age in years	0–	3–	6–	9–	12–15
Number of children	19	12	17	15	13

(a) Calculate approximate values for the mean and standard deviation of the ages of these 76 children. Give your answers to 1 decimal place.

(b) Six months later the same 76 children are living on the Millhill estate. Write down values for the mean and standard deviation of their ages at this time.

5 The minimum temperature on a night in May was measured at a random sample of 21 places on the south coast of England. The measurements, in °Celsius, are shown below.

Temperature (°C)	7	8	9	10	11	12
Frequency	1	3	6	5	4	2

(a) Calculate estimates of the mean and standard deviation of minimum temperatures on the south coast of England that night. Give your answers to 1 d.p.

(b) Temperatures in °Celsius can be converted to °Fahrenheit by multiplying by 1.8 and adding 32. Find estimates for the mean and standard deviation of minimum temperatures in °Fahrenheit on the south coast of England that night.

6 Two varieties of apple are grown in an orchard. The table shows the weights of random samples of 36 apples of variety A and 36 apples of variety B.

Weight (g)	Variety A	Variety B
70–	3	8
80–	10	7
85–	14	7
90–	6	4
95–	3	5
100–	0	3
110–120	0	2

(a) Find estimates of the mean and standard deviation of weights for each of the two varieties of apple.

(b) Compare the distribution of weights for the two varieties.

Test yourself

What to review

If your answer is incorrect:

1 Find the mode, median, range and interquartile range for the following set of numbers.
11 19 9 15 16 7 18 11 14 9 15 19 9 13

See p 2 Example 1 or review Advancing Maths for AQA S1 pp 9, 17

2 A ferry carries passengers between an island and the mainland. One day the ferry makes 12 crossings. The number of passengers for each journey from the mainland to the island is given below.
6 13 20 15 22 19 20 14 10 8 9 3

Calculate the mean and standard deviation of the number of passengers per journey from the mainland to the island on.

See p 2 Example 1 or review Advancing Maths for AQA S1 pp 9, 20

3 A baker makes chocolate cookies. The distribution of weights, to the nearest gram, of a random sample of 52 cookies is shown below.

Weight (g)	56	57	58	59	60	61	62
Frequency	4	7	11	13	10	5	2

(a) Calculate estimates of:
 (i) the mean weight of cookies made by the baker, giving your answer to the nearest gram,
 (ii) the standard deviation of the weights of cookies made by the baker, giving your answer in grams to 1 d.p.

(b) Show that the weight of the heaviest cookie is within three standard deviations of the mean.

See p 3 Example 2 or review Advancing Maths for AQA S1 pp 9–10 and 22–23

4 The speeds of a random sample of cars travelling along a stretch of a road on a fine day are shown in the table.

Speed (mph)	35–	45–	50–	55–	60–	65–70
Frequency	7	28	43	51	38	17

(a) Find estimates of the mean and standard deviation of speeds.

(b) On a foggy day the estimated mean and standard deviation of speeds of cars were 38.4 mph and 1.06 mph respectively. Compare the travelling speeds on this road for a fine day and a foggy day.

See p 4 Example 3 or review Advancing Maths for AQA S1 pp 13–14, 23–24 and 28

5 The weekly wages of workers in a small factory have mean £260 and standard deviation £12.60. Find the mean and standard deviation of wages if:

(a) every worker's wage is increased by £10,

(b) every worker's wage is increased by 5%.

See p 3 Example 2 or review Advancing Maths for AQA S1 p 26

Test yourself ANSWERS

1 Mode 9, median 13.5, range 12, interquartile range 7

2 Mean 13.3; standard deviation 6.17 (5.90 accepted)

3 (a) (i) 59 g, (ii) 1.5 g

(b) Mean + (3 × standard deviation) = 63.4 g; heaviest cookie weighs less than this.

4 (a) Mean 56.1 mph, standard deviation 6.74 mph (6.72 accepted)

(b) Speeds are much slower on average on the foggy day and the low standard deviation shows that all vehicles travel at roughly the same speed.

5 (a) Mean £270, standard deviation £12.60

(b) Mean £273, standard deviation £13.23

CHAPTER 2

Probability

Key points to remember

1 Probability is measured on a scale from 0 to 1.

2 If a trial can result in one of n equally likely outcomes and an event consists of r of these, then the probability of the event happening as a result of the trial is $\frac{r}{n}$.

3 Two events are **mutually exclusive** if they cannot both happen.

4 If A and B are mutually exclusive events $P(A \cup B) = P(A) + P(B)$.

5 The event of A not happening as the result of a trial is called the **complement** of A and is usually denoted A'.

6 Two events are **independent** if the probability of one happening is unaffected by whether or not the other happens.

7 If A and B are independent events $P(A \cap B) = P(A)P(B)$.

8 $P(A|B)$ denotes the probability that event A happens given that event B happens.

9 If events A and B are independent $P(A|B) = P(A)$.

10 $P(A \cup B) = P(A) + P(B) - P(A \cap B)$

11 $P(A \cap B) = P(A)P(B|A) = P(B)P(A|B)$

Worked example 1

A box contains 24 sweets. Eight of them are flavoured lemon, six are flavoured lime and the rest are flavoured orange. Marie picks a sweet at random. Find the probability that it is:

(a) orange flavoured,

(b) either orange or lemon flavoured,

(c) not lemon flavoured.

The 24 sweets are all equally likely to be picked.

(a) Number of orange sweets = $24 - (8 + 6) = 10$.

$P(\text{orange}) = \frac{10}{24} = \frac{5}{12}$

Using **2**

(b) P(orange or lemon) $= \dfrac{10 + 8}{24} = \dfrac{18}{24} = \dfrac{3}{4}$

(c) The number of sweets not lemon flavoured $= 24 - 8 = 16$.

P(not lemon flavoured) $= \dfrac{16}{24} = \dfrac{2}{3}$

2

Worked example 2

Ann, Bharti and Chloe travel to school independently of each other. The probabilities that the girls are late for school are 0.2, 0.1 and 0.3 respectively. Find the probability that, on a particular day:

(a) Chloe is not late for school,

(b) Ann and Bharti are both late,

(c) Ann and Bharti are late but Chloe is on time.

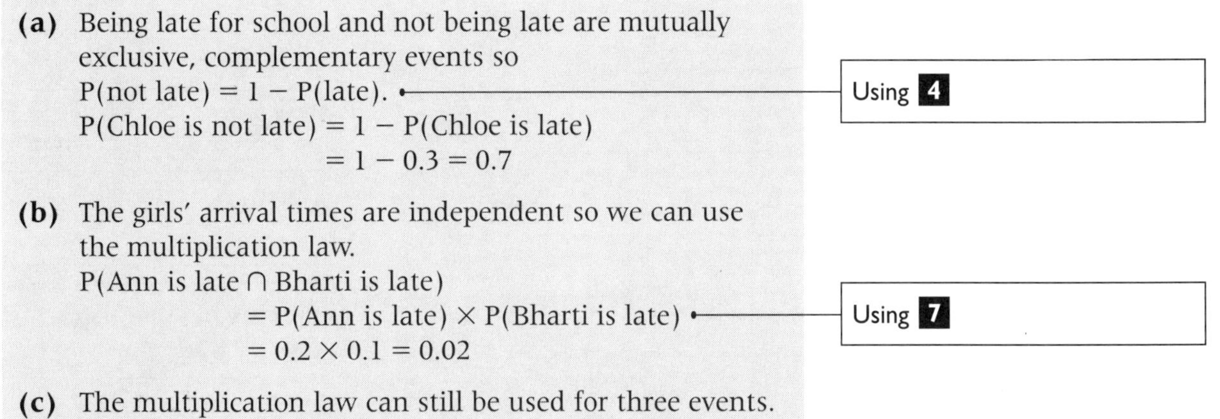

(a) Being late for school and not being late are mutually exclusive, complementary events so
P(not late) = 1 − P(late).
P(Chloe is not late) = 1 − P(Chloe is late)
$= 1 - 0.3 = 0.7$

Using **4**

(b) The girls' arrival times are independent so we can use the multiplication law.
P(Ann is late ∩ Bharti is late)
= P(Ann is late) × P(Bharti is late)
$= 0.2 \times 0.1 = 0.02$

Using **7**

(c) The multiplication law can still be used for three events.
P(Ann is late ∩ Bharti is late ∩ Chloe is not late)
$= 0.2 \times 0.1 \times 0.7 = 0.014$

Worked example 3

A small primary school has three classes. The number of boys and girls in each class are shown below.

	Class 1	Class 2	Class 3
Number of boys	5	8	12
Number of girls	10	8	8

The school raises money for a local charity and the head teacher chooses three pupils at random, one from each class, to present the cheque to the charity.
Find the probability that she chooses:

(a) three girls,

(b) three children of the same sex,

(c) two boys and one girl.

A tree diagram can be useful to sort out the probabilities. Three children are to be chosen so the diagram needs three sets of branches to show the probabilities of choosing a boy or a girl from each class.

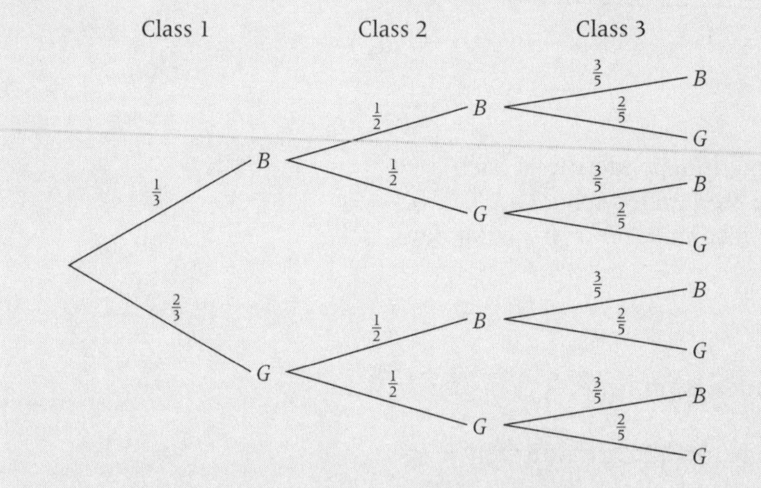

Remember the tree diagram is just to help you work things out. It doesn't need to be a work of art so don't spend too long on it.

(a) The bottom route through the tree is required.

$$P(G \cap G \cap G) = \frac{2}{3} \times \frac{1}{2} \times \frac{2}{5} = \frac{4}{30} = \frac{2}{15}$$

Multiply probabilities when you follow a route through the tree.

(b) Either three boys or three girls would satisfy the requirement.

$$P(B \cap B \cap B) + P(G \cap G \cap G) = \left(\frac{1}{3} \times \frac{1}{2} \times \frac{3}{5}\right) + \frac{4}{30}$$
$$= \frac{3}{30} + \frac{4}{30} = \frac{7}{30}$$

Add probabilities to combine the results from different routes through the tree.

(c) Three roots through the tree satisfy the requirement.

$$P(B \cap B \cap G) + P(B \cap G \cap B) + P(G \cap B \cap B)$$
$$= \left(\frac{1}{3} \times \frac{1}{2} \times \frac{2}{5}\right) + \left(\frac{1}{3} \times \frac{1}{2} \times \frac{3}{5}\right) + \left(\frac{2}{3} \times \frac{1}{2} \times \frac{3}{5}\right)$$
$$= \frac{2}{30} + \frac{3}{30} + \frac{6}{30} = \frac{11}{30}$$

It makes the working easier if you don't do any cancelling here. Notice that all the denominators are 30 which makes the addition simple.

Worked example 4

A Business College offers full-time one-year and two-year courses and various part-time courses. The following table shows the number of students enrolled on each type of course, classified by age group.

	Full-time one-year	Full-time two-year	Part-time
Aged 25 or under	65	20	20
Aged over 25	35	15	45

A student is chosen at random from the 200 students enrolled.
S is the event that the chosen student is aged 25 or under.
T is the event that the chosen student is on a part-time course.
T' is the event not T.

(a) Find the value of:

(i) $P(S)$,

(ii) $P(S|T)$,

(iii) $P(T'|S)$.

(b) Two different students are chosen at random from those enrolled on a full-time two-year course. Find the probability that they are both aged over 25.

(a) (i) We are concerned with all 200 students. The number aged 25 or under is 65 + 20 + 20 = 105.

$$P(S) = \frac{105}{200} = 0.525$$

(ii) We are concerned only with students on part-time courses. There are 65 of these and 20 are aged 25 or under.

$$P(S|T) = \frac{20}{65} = 0.308$$

Keep referring back to the table to find the correct numbers.

(iii) We are concerned only with the 105 students aged 25 or under. T′ is the event that a student is not on a part-time course, that is, the student is on a full-time course. There are 85 such students out of the 105 aged 25 or under.

$$P(T'|S) = \frac{85}{105} = 0.810$$

(b) This is selection without replacement. We need to use the multiplication law, but the probabilities change each time a student is selected.

When choosing the first student there are 35 students on full-time two-year courses of whom 15 are aged over 25.

$$P(\text{first student over 25}) = \frac{15}{35}$$

When the second choice is made, one student aged over 25 has been removed so there are now 34 students to choose from of whom 14 are aged over 25.

$$P(\text{second student over 25}|\text{first student over 25}) = \frac{14}{34}$$

$$P(\text{both aged over 25}) = \frac{15}{35} \times \frac{14}{34} = \frac{210}{1190} = 0.176$$

Using **11**

Worked example 5

A card is drawn at random from a pack of 52 playing cards.
X is the event that the card is a club.
Y is the event that the card is a picture card.

(a) Show that events X and Y are independent.

(b) Find $P(X \cup Y)$.

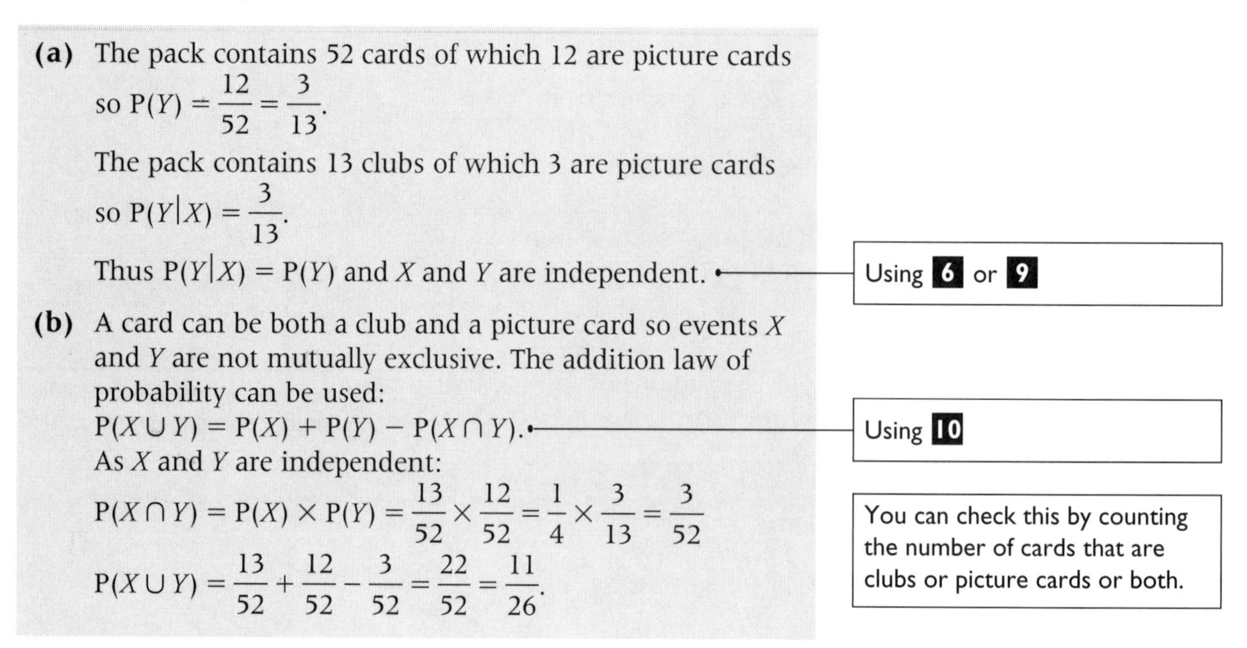

(a) The pack contains 52 cards of which 12 are picture cards so $P(Y) = \frac{12}{52} = \frac{3}{13}$.
The pack contains 13 clubs of which 3 are picture cards so $P(Y|X) = \frac{3}{13}$.
Thus $P(Y|X) = P(Y)$ and X and Y are independent.

Using **6** or **9**

(b) A card can be both a club and a picture card so events X and Y are not mutually exclusive. The addition law of probability can be used:
$P(X \cup Y) = P(X) + P(Y) - P(X \cap Y)$.

Using **10**

As X and Y are independent:
$$P(X \cap Y) = P(X) \times P(Y) = \frac{13}{52} \times \frac{12}{52} = \frac{1}{4} \times \frac{3}{13} = \frac{3}{52}$$
$$P(X \cup Y) = \frac{13}{52} + \frac{12}{52} - \frac{3}{52} = \frac{22}{52} = \frac{11}{26}.$$

You can check this by counting the number of cards that are clubs or picture cards or both.

REVISION EXERCISE 2

1 The 12 picture cards are taken out of a pack of playing cards. One of these 12 cards is chosen at random. Find the probability that it is:

(a) a king,

(b) a diamond,

(c) the king of diamonds.

2 Jane's bus to work is scheduled to leave at 8.10 am. The probability that it leaves after 8.10 am but before 8.15 am is 0.55. The probability that it is 5 or more minutes late is 0.12. Find the probability that Jane's bus leaves:

(a) late,

(b) on time.

3 Mandy and Sue often play a game of squash after work. The probability that Mandy wins a game is 0.6.
For a series of three games, find the probability that:

(a) Sue wins all three games,

(b) Sue wins at least one game.

4 A tetrahedral die has four faces numbered 1 to 4. When it is thrown onto a table, the score is the number on the face that rests on the table.
Two such dice are thrown together. Find the probability that:

(a) both dice score 3,

(b) at least one of the dice scores 3,

(c) both dice score the same number,

(d) the sum of the two scores is 6.

5 Amin either cycles to school or goes by bus. His decision about how to travel depends on the weather. The following table shows the probabilities that he cycles to school for various types of weather.

	Warm	**Cold**
Wet	0.5	0.2
Dry	0.9	0.7

(a) One week in June it is warm and dry for three consecutive days. Find the probability that Amin cycles to school on all three days.

(b) A Monday in September is warm and dry; the next day is warm and wet. Find the probability that Amin goes to school by bus on both days.

(c) A Thursday and Friday in November are both cold and dry. Find the probability that Amin travels to school by the same means on both days.

6 During December it is cold every day. If a day is wet, there is a probability of 0.7 that the next day will also be wet. If a day is dry, there is a probability of 0.8 that the next day will also be dry.

(a) Monday, December 6th is wet. Find the probability that Wednesday, December 8th is: (i) wet, (ii) dry.

(b) The probabilities for Amin's methods of travelling to school are those given in question **5**. Find the probability that Amin cycles to school on Wednesday, December 8th.

7 A badminton club has 36 members. The table shows membership numbers classified by age and gender.

	Under 20	**20–35**	**Over 35**
Female	4	8	2
Male	6	12	4

The club is given two tickets to a national tournament. It is decided to choose one member at random to receive the tickets.
A is the event that the member chosen is female.
B is the event that the member chosen is under 20 years old.
C is the event that the member chosen is over 35 years old.
B' is the event not B.

Find:

(a) $\mathrm{P}(A \cap B)$, **(b)** $\mathrm{P}(A \cup B')$, **(c)** $\mathrm{P}(B' | A)$, **(d)** $\mathrm{P}(C | (A \cap B'))$.

Test yourself

What to review

If your answer is incorrect:

1 A box contains 20 coloured balls of which 4 are red, 8 green, 6 blue and the rest yellow. A ball is chosen at random from the box. Find the probability that it is:
(a) green,
(b) either red or blue,
(c) not yellow.

See p 8 Example 1 or review Advancing Maths for AQA S1 p 34

---

2 Contestants in a quiz show are offered four possible answers to each question. Simon guesses the answers to three of the questions. Find the probability that he guesses:
(a) all three answers correctly,
(b) all three answers incorrectly,
(c) one answer correctly and two answers incorrectly.

See p 9 Examples 2 and 3 or review Advancing Maths for AQA S1 pp 39–41

---

3 In a charity shop a box contains 30 assorted books. The assortment includes 12 crime fiction and 6 romantic fiction books. The rest of the books are non-fiction.
A customer picks out two books at random.
Find the probability that:
(a) she chooses two romantic fiction books,
(b) neither book is non-fiction,
(c) she chooses one fiction and one non-fiction book.

see pp 9–10 Examples 3 and 4(b) or review Advancing Maths for AQA S1 pp 48–49

---

4 A college offers Easter revision courses for AS-level and A-level science students. On a day when the college is inspected, the numbers of students taking the available courses are as shown in the table.

	Physics	**Chemistry**	**Biology**
AS-level	15	15	20
A-level	10	8	12

The inspector chooses one student at random to interview.
M is the event that an AS-level student is chosen.
R is the event that a Physics student is chosen.
S is the event that a Biology student is chosen.
R' is the event not R.
S' is the event not S.
(a) Name two events that are mutually exclusive.
(b) Show that events M and S are independent.
(c) Find: (i) $P(M \cap R)$, (ii) $P(M \cup S')$, (iii) $P(R' | M)$.
(d) State in words the event $R' \cap S'$.

See pp 10–12 Examples 4 and 5 or review Advancing Maths for AQA S1 pp 36, 38 and 43–45

Test yourself ANSWERS

1 **(a)** $\frac{2}{5}$ **(b)** $\frac{1}{2}$ **(c)** $\frac{9}{10}$

2 **(a)** $\frac{1}{64}$ **(b)** $\frac{27}{64}$ **(c)** $\frac{27}{64}$

3 **(a)** 0.0345 **(b)** 0.352 **(c)** 0.497

4 **(a)** R and S

(b) $P(M) = \frac{50}{80} = \frac{5}{8}$, $P(M|S) = \frac{20}{32} = \frac{5}{8}$

(c) (i) $\frac{3}{16}$ (ii) $\frac{17}{20}$ (iii) $\frac{7}{10}$

(d) A Chemistry student is chosen.

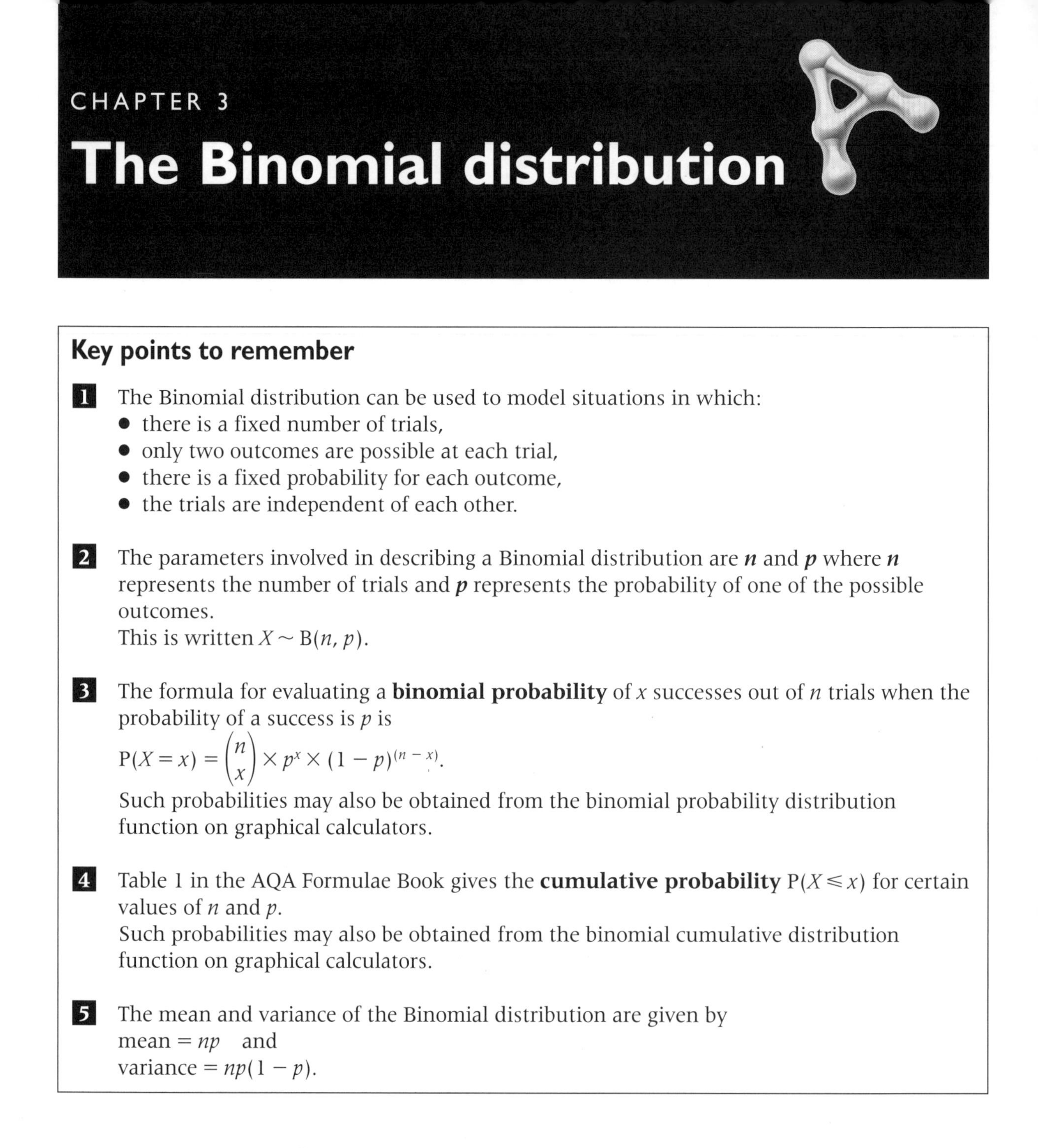

CHAPTER 3

The Binomial distribution

Key points to remember

1 The Binomial distribution can be used to model situations in which:
- there is a fixed number of trials,
- only two outcomes are possible at each trial,
- there is a fixed probability for each outcome,
- the trials are independent of each other.

2 The parameters involved in describing a Binomial distribution are $\boldsymbol{n}$ and $\boldsymbol{p}$ where $\boldsymbol{n}$ represents the number of trials and $\boldsymbol{p}$ represents the probability of one of the possible outcomes.
This is written $X \sim \mathrm{B}(n, p)$.

3 The formula for evaluating a **binomial probability** of x successes out of n trials when the probability of a success is p is
$\mathrm{P}(X = x) = \binom{n}{x} \times p^x \times (1 - p)^{(n - x)}$.
Such probabilities may also be obtained from the binomial probability distribution function on graphical calculators.

4 Table 1 in the AQA Formulae Book gives the **cumulative probability** $\mathrm{P}(X \leqslant x)$ for certain values of n and p.
Such probabilities may also be obtained from the binomial cumulative distribution function on graphical calculators.

5 The mean and variance of the Binomial distribution are given by
mean $= np$ and
variance $= np(1 - p)$.

Worked example 1

A college student travels to college by train. The probability that the train is late is 0.2. Over the next three weeks, out of a total of 15 journeys, find the probability that the train is late:

(a) five or fewer times,

(b) at most six times,

(c) more than 5 times,

(d) between 6 and 8 (inclusive) times,

(e) fewer than 4 times.

(f) Find the mean and the standard deviation of the number of times in three weeks that the student finds the train is late.

(a) $n = 15, p = 0.2, x = 5$ — $\overline{0\,1\,2\,3\,4\,5}|\,6\,7\ldots$
$P(X \leqslant 5) = 0.9389$ — Using **4**

(b) $n = 15, p = 0.2, x = 6$ — $\overline{0\,1\,2\,3\,4\,5\,6}|\,7\ldots$
$P(X \leqslant 6) = 0.9819$ — Using **4**

(c) $n = 15, p = 0.2, x = 5$ — $0\,1\,2\,3\,4\,5\,|\overline{6\,7}\ldots$
$P(X \leqslant 5) = 0.9389$ so $P(X > 5) = 1 - 0.9389 = 0.0611$ — Using **4**

(d) $P(6 \leqslant X \leqslant 8) = P(X \leqslant 8) - P(X \leqslant 5)$ — $\overline{0\,1\,2\,3\,4\,5}|\,6\,7\,8|\,9\ldots$
Using $n = 15, p = 0.2$ with $x = 8$ and $x = 5$
gives $0.9992 - 0.9389 = 0.0603$. — Using **4**

(e) $P(X < 4) = P(X \leqslant 3)$ — $\overline{0\,1\,2\,3}|\,4\,5\,6\,7\ldots$
Using $n = 15, p = 0.2$ and $x = 3$
gives 0.6482. — Using **4**

(f) mean $= np = 15 \times 0.2 = 3$
variance $= np(1 - p) = 15 \times 0.2 \times 0.8 = 2.4$ — Using **5**
standard deviation $= \sqrt{2.4} = 1.55$

Worked example 2

The probability of a light fitting being damaged in delivery is 0.3. Find the probability that, in a randomly selected sample of 50 light fittings, the number damaged is:

(a) at most 12, **(b)** exactly 15,

(c) 20 or more,

(d) between 10 and 15 (inclusive).

(e) Find the mean and standard deviation of the number of damaged light fittings in a sample of 50.

(a) $n = 50, p = 0.3, x = 12$ — $\overline{0\,1\ldots10\,11\,12}|\,13\ldots$
$P(X \leqslant 12) = 0.2229$ — Using **4**

(b) This requires use of the formula
$P(X = x) = \binom{n}{x} \times p^x \times (1 - p)^{(n - x)}$, — Using **3**
with $n = 50, p = 0.3$ and $x = 15$,
or Table 1 using $P(X \leqslant 15) - P(X \leqslant 14)$
$= 0.5692 - 0.4468 = 0.1224$. — Using **4**

The formula gives $P(X = 15) = \binom{50}{15} \times 0.3^{15} \times 0.7^{35}$
$= 0.122$ (3 s.f.).

$\binom{50}{15} = {}^{50}C_{15} = \dfrac{50!}{15! \times 35!}$

3

(c) $P(X \geqslant 20) = 1 - P(X \leqslant 19)$ 1 … 18 19 | 20 21 …
Using $n = 50$, $p = 0.3$ and $x = 19$
gives $P(X \leqslant 19) = 0.9152$,
so $P(X \geqslant 20) = 1 - 0.9152 = 0.0848$. — Using **4**

(d) $P(10 \leqslant X \leqslant 15) = P(X \leqslant 15) - P(X \leqslant 9)$
0 1 … 9 | 10 11 12 13 14 15 | …
Using $n = 50$, $p = 0.3$ with $x = 15$ and $x = 9$,
gives $0.5692 - 0.0402 = 0.529$. — Using **4**

(e) mean $= np = 50 \times 0.3 = 15$
variance $= np(1 - p) = 50 \times 0.3 \times 0.7 = 10.5$ — Using **5**
standard deviation $= \sqrt{10.5} = 3.24$

Worked example 3

Tasmeen, a trainee holiday advisor, calculates the cost of package holidays to Spain using a computer package. The cost of such holidays varies according to several different circumstances. During her first week at work, Tasmeen has a probability of 0.25 of calculating the cost of a holiday incorrectly.

(a) Given that Tasmeen calculates the cost of 22 holidays during her first week at work, find the probability that exactly two are incorrectly calculated.

(b) Given that Tasmeen calculates the cost of 30 holidays during her second week at work, and that she now has a probability of 0.2 of calculating the cost of a holiday incorrectly, find the probability that:
(i) fewer than five are incorrectly calculated,
(ii) more than 26 are correctly calculated.

(c) The random variable, C, represents the number of holiday costs Tasmeen calculates until a total of eight have been calculated correctly. State, giving a reason, whether the Binomial distribution is an appropriate model for C.

(d) A random sample of 50 holiday costs are taken from those calculated by Tasmeen during her first 6 months of employment. Give a reason why the Binomial distribution may not provide a suitable model for the number of incorrect costs in this sample.

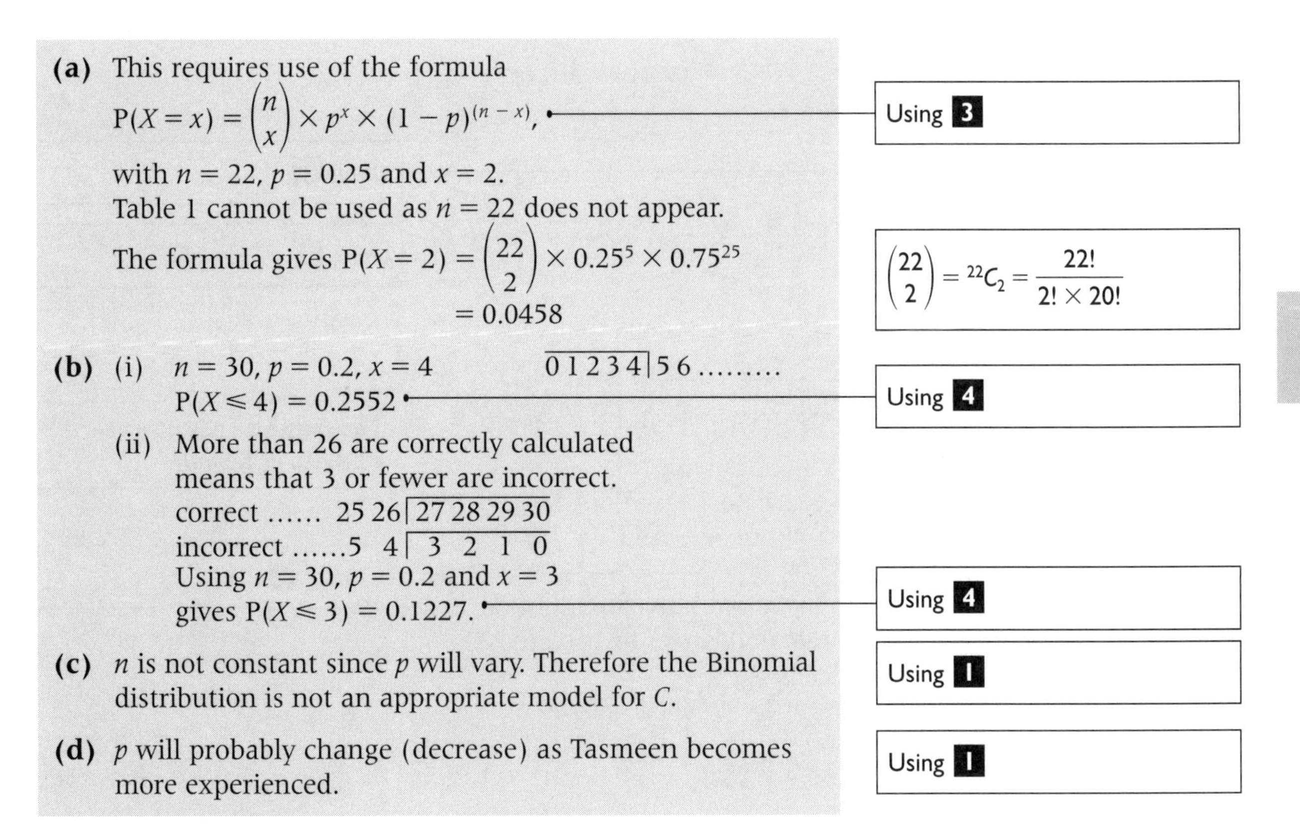

(a) This requires use of the formula

$P(X = x) = \binom{n}{x} \times p^x \times (1 - p)^{(n-x)}$, — Using **3**

with $n = 22$, $p = 0.25$ and $x = 2$.

Table 1 cannot be used as $n = 22$ does not appear.

The formula gives $P(X = 2) = \binom{22}{2} \times 0.25^5 \times 0.75^{25}$

$= 0.0458$

$\binom{22}{2} = {}^{22}C_2 = \frac{22!}{2! \times 20!}$

(b) (i) $n = 30$, $p = 0.2$, $x = 4$ $\overline{0\,1\,2\,3\,4}|\,5\,6\,\ldots\ldots\ldots$

$P(X \leqslant 4) = 0.2552$ — Using **4**

(ii) More than 26 are correctly calculated means that 3 or fewer are incorrect.

correct …… 25 26 | 27 28 29 30

incorrect ……5 4 | 3 2 1 0

Using $n = 30$, $p = 0.2$ and $x = 3$

gives $P(X \leqslant 3) = 0.1227$. — Using **4**

(c) n is not constant since p will vary. Therefore the Binomial distribution is not an appropriate model for C. — Using **1**

(d) p will probably change (decrease) as Tasmeen becomes more experienced. — Using **1**

3

Worked example 4

Each house in a local authority area is supplied with a recycling box. The council arranges a weekly collection of these boxes from outside the houses.

In a street containing 40 houses, the number of boxes placed outside for a particular collection is a random variable X. The values of X for 10 consecutive collections were as below.

23 20 25 23 24 26 27 22 24 26

(a) Calculate the mean and standard deviation of the given data.

(b) Use the data to estimate p, the proportion of houses that will have a box placed outside for a particular collection.

(c) Suggest the name of a distribution which might be a suitable model for the random variable X.

(d) Use the distribution you have suggested in part **(c)** and the value of p you estimated in part **(b)** to find:

(i) the probability that X is greater than 25,

(ii) the probability that X is exactly 23,

(iii) the mean and the standard deviation of X.

(e) By comparing your answers to part **(a)** with those to part **(d)** (iii), explain whether the distribution you suggested in part **(c)** provides a suitable model.

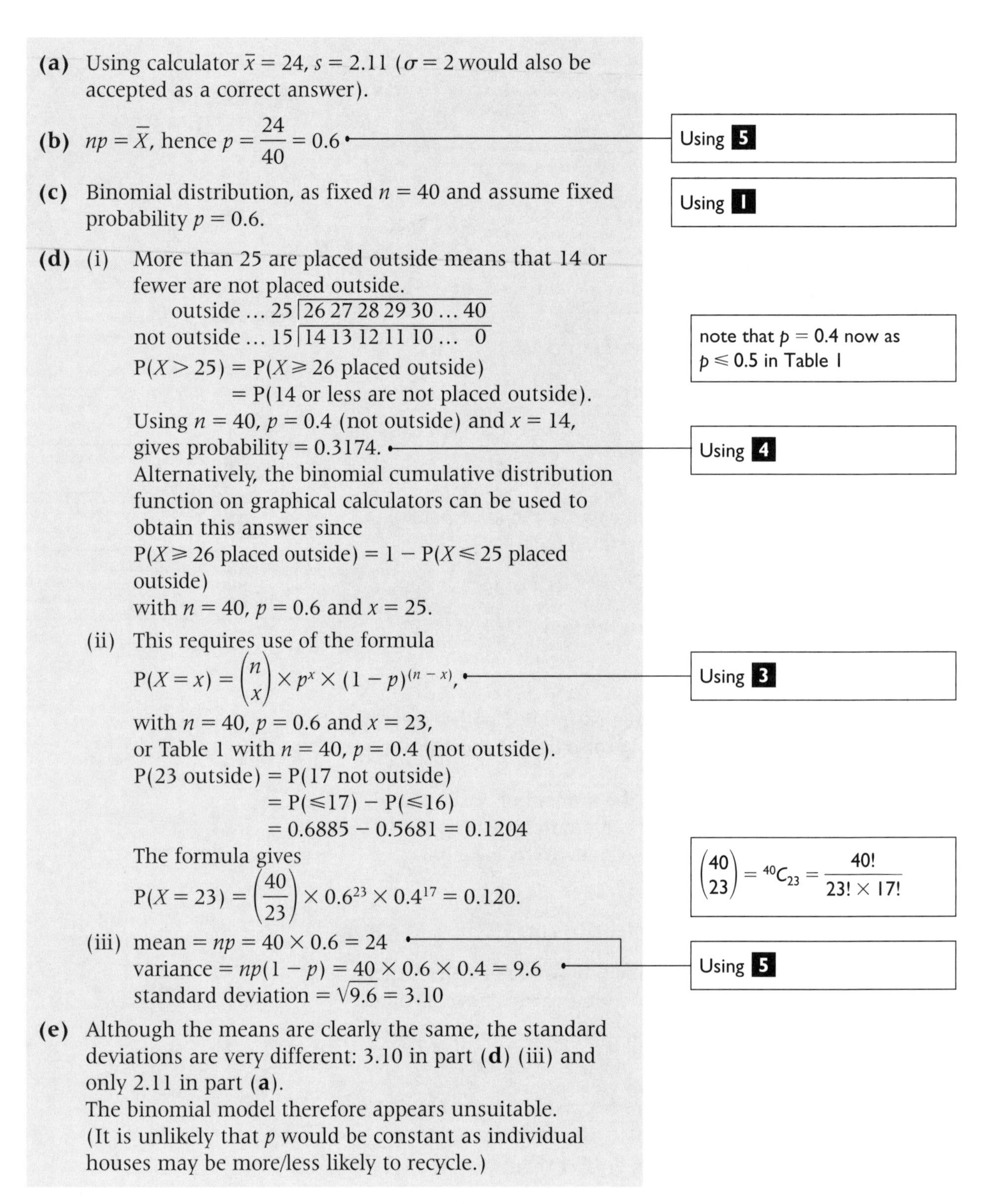

(a) Using calculator $\bar{x} = 24$, $s = 2.11$ ($\sigma = 2$ would also be accepted as a correct answer).

(b) $np = \bar{X}$, hence $p = \dfrac{24}{40} = 0.6$ — Using **5**

(c) Binomial distribution, as fixed $n = 40$ and assume fixed probability $p = 0.6$. — Using **1**

(d) (i) More than 25 are placed outside means that 14 or fewer are not placed outside.

outside … 25 | 26 27 28 29 30 … 40

not outside … 15 | 14 13 12 11 10 … 0

$P(X > 25) = P(X \geqslant 26 \text{ placed outside})$

$= P(14 \text{ or less are not placed outside})$.

Using $n = 40$, $p = 0.4$ (not outside) and $x = 14$, gives probability $= 0.3174$. — Using **4**

note that $p = 0.4$ now as $p \leqslant 0.5$ in Table 1

Alternatively, the binomial cumulative distribution function on graphical calculators can be used to obtain this answer since

$P(X \geqslant 26 \text{ placed outside}) = 1 - P(X \leqslant 25 \text{ placed outside})$

with $n = 40$, $p = 0.6$ and $x = 25$.

(ii) This requires use of the formula

$P(X = x) = \binom{n}{x} \times p^x \times (1 - p)^{(n - x)}$, — Using **3**

with $n = 40$, $p = 0.6$ and $x = 23$,

or Table 1 with $n = 40$, $p = 0.4$ (not outside).

$P(23 \text{ outside}) = P(17 \text{ not outside})$

$= P(\leqslant 17) - P(\leqslant 16)$

$= 0.6885 - 0.5681 = 0.1204$

The formula gives

$P(X = 23) = \binom{40}{23} \times 0.6^{23} \times 0.4^{17} = 0.120.$

$\binom{40}{23} = {}^{40}C_{23} = \dfrac{40!}{23! \times 17!}$

(iii) mean $= np = 40 \times 0.6 = 24$

variance $= np(1 - p) = 40 \times 0.6 \times 0.4 = 9.6$ — Using **5**

standard deviation $= \sqrt{9.6} = 3.10$

(e) Although the means are clearly the same, the standard deviations are very different: 3.10 in part **(d)** (iii) and only 2.11 in part **(a)**.

The binomial model therefore appears unsuitable.

(It is unlikely that p would be constant as individual houses may be more/less likely to recycle.)

REVISION EXERCISE 3

1 A commuter takes the bus into an office in town every day. The probability that the bus is late is 0.3. Over the next two weeks, out of a total of 10 journeys, find the probability that the bus is late:

(a) three or fewer times,

(b) at most twice,

(c) more than 4 times,

(d) between 2 and 4 (inclusive) times,

(d) fewer than 4 times.

(e) Find the mean and the standard deviation of the number of times in two weeks that the bus is late.

2 A golfer practises on a driving range. The objective is to drive a ball to within 15 m of a flag.

(a) On the first visit, the independent probability of success with each particular drive is 0.25.
If 12 balls are driven, find the probability of:

(i) 4 or fewer successes,

(ii) 4 or more successes.

(b) Some weeks later, the independent probability of success has increased to 0.26. Find the probability of 5 successes in 25 drives.

3 A doctor knows from experience that 15% of patients who are given a certain medicine will have undesirable side-effects. Find the probability that, among 15 randomly selected patients:

(a) (i) none will have side-effects,

(ii) exactly 3 will have side-effects,

(iii) at least 2 will have side-effects.

An ointment, also used with the doctor's patients, leads to undesirable side effects in 35% of patients. Find the probability that, among 40 randomly selected patients:

(b) (i) at most 12 will have side-effects,

(ii) between 10 and 20 will have side effects,

(iii) more than half will have side-effects.

4 An examination has 40 multiple-choice questions, each with five different alternative answers, only one of which is correct and is allotted one mark. Assume the response to each question is independent of the response to any other question. No marks are awarded for incorrect answers.
For a candidate who guesses the answers to all forty questions, find the probability that:

(a) (i) the candidate obtains at least 12 marks,

(ii) the candidate obtains more than 10 marks.

(b) Find the mean and the standard deviation of the number of correct answers out of the 40 questions for a candidate who guesses the answers.

5 Eight friends take a picnic to a cricket match. As her contribution to the picnic, Hilda buys eight sandwiches at a supermarket. She selects the sandwiches at random from those on display. The probability that a sandwich is suitable for vegetarians is independently 0.3 for each sandwich.

(a) Find the probability that, of the eight sandwiches, the number suitable for vegetarians is:
(i) 2 or fewer,
(ii) exactly two,
(iii) more than 3.

(b) Two of the friends are vegetarians. Hilda decides to ensure that the eight sandwiches she takes to the match will include at least two suitable for vegetarians. If, having selected eight sandwiches at random, she finds that they include fewer than two suitable for vegetarians, then she will replace one, or if necessary two, of the sandwiches unsuitable for vegetarians with the appropriate number of sandwiches suitable for vegetarians.

State whether or not the Binomial distribution provides an adequate model for the number of sandwiches suitable for vegetarians which Hilda takes to the match. Explain your answer.

6 A campaign to combat the economic devastation caused to car manufacturing communities by factory closures employed a researcher. The campaign organisers wished to know the proportion of redundant car factory workers who were able to find alternative employment within a year of being made redundant.

The researcher found that the probability of a redundant car factory worker visited at home refusing to answer a questionnaire was, independently, 0.15.

What is the probability, on a day when 12 redundant car factory workers are visited at home:

(a) (i) fewer than 3 will refuse to answer the questionnaire,
(ii) exactly two will refuse to answer the questionnaire,
(iii) at least nine will agree to answer the questionnaire.

In order to save time, a postal questionnaire is sent to a random sample of 30 redundant car factory workers. It is known that the probability that a postal questionnaire will be returned is, independently, 0.2. Find the probability that, of the 30 questionnaires:

(b) (i) no more than five are returned,
(ii) more than six are returned,
(iii) exactly seven are returned.

7 Richard is trying to sell raffle tickets in order to raise money for a charity. The probability that he makes a sale is, independently, 0.08 for each person he asks. Given that he asks 50 people, find the probability that he will sell:

(a) (i) 3 or fewer tickets,
(ii) more than 6 tickets,
(iii) between 5 and 10 tickets,
(iv) exactly 4 tickets.

(b) Find the mean amount of money that Richard will raise for the charity when he asks 50 people, if each ticket costs £2.00.

(c) Give one reason why the Binomial distribution might not provide an adequate model for the number of tickets sold by Richard.

8 Each evening Aaron sets his alarm for 7 am. He believes the probability of him waking before his alarm rings each morning is 0.4 and is independent from morning to morning.

(a) Assuming that Aaron's belief is correct, determine the probability that, during a week (7 mornings), he wakes before his alarm rings:
(i) on 2 or fewer mornings,
(ii) on more than 1 but fewer than 5 mornings.

(b) Assuming that Aaron's belief is correct, calculate the probability that, during a 4-week period, he wakes before the alarm rings on exactly 7 mornings.

(c) Assuming that Aaron's belief is correct, calculate values for the mean and standard deviation of mornings in a week when Aaron wakes before his alarm rings.

(d) During a 50-week period, Aaron records, each week, the number of mornings on which he wakes before his alarm rings. The results are as follows.

Number of mornings	0	1	2	3	4	5	6	7
Frequency	10	8	7	7	5	5	4	4

(i) Calculate the mean and standard deviation of these data.
(ii) State, giving reasons, whether your answers to part **(d)** (i) support Aaron's belief that the probability of him waking before the alarm rings each morning is 0.4 and is independent from morning to morning.

Test yourself	What to review
	If your answer is incorrect:
1 The probability that a shopper will purchase the detergent that is on offer at a supermarket is 0.65. Find the probability that exactly 5, out of a random selection of 8 shoppers who purchased detergent, bought the detergent that was on offer.	See p 17 Example 2 or review Advancing Maths for AQA S1 pp 60–64
2 Rods produced at a factory are tested in batches of 50. If the probability that a rod is damaged in the manufacturing process is 0.09, find the probability that, in a randomly selected batch, fewer than 5 are damaged.	See p 16 Example 1 or review Advancing Maths for AQA S1 pp 65–68
3 Of the pupils at a school, 30% travel to school by bus. If a random sample of 20 pupils is selected, find the probability that: **(a)** at least 5 travel by bus, **(b)** at most 4 travel by bus, **(c)** more than 2 travel by bus, **(d)** at least 14 do not travel by bus. **(e)** Find the mean and standard deviation of the number of pupils in a sample of 20 who travel by bus.	See p 16 Example 1 or review Advancing Maths for AQA S1 pp 65–69
4 Parijat walks to his office every morning. The probability that he arrives late is 0.15. Assume this probability is independent of whether he arrives late on any other morning. Find, for a week when he walks to his office on five mornings, the probability that: **(a)** he arrives late on two or fewer mornings, **(b)** he arrives late on more than three mornings. **(c)** Find the mean and standard deviation of the number of mornings on which he arrives late. **(d)** In the context of this question, give two possible reasons why the binomial distribution may not be a suitable model for the number of mornings Parijat arrives late in a week.	See pp 17–19 Examples 2 and 3 or review Advancing Maths for AQA S1 pp 65–69 and 72–75

Test yourself ANSWERS

1 0.279

2 0.528

3 **(a)** 0.7625 **(b)** 0.2375 **(c)** 0.9645
(d) $P(X \leqslant 6) = 0.608$ **(e)** $\mu = 6, \sigma = 2.05$

4 **(a)** 0.973
(b) 0.002 23
(c) $\mu = 0.75, \sigma = 0.798$
(d) Probability he arrives late may not be constant (may vary according to the weather, for example).
Probability he arrives late may not be independent (if he is late on Monday he may make more effort to be on time on Tuesday, for example).

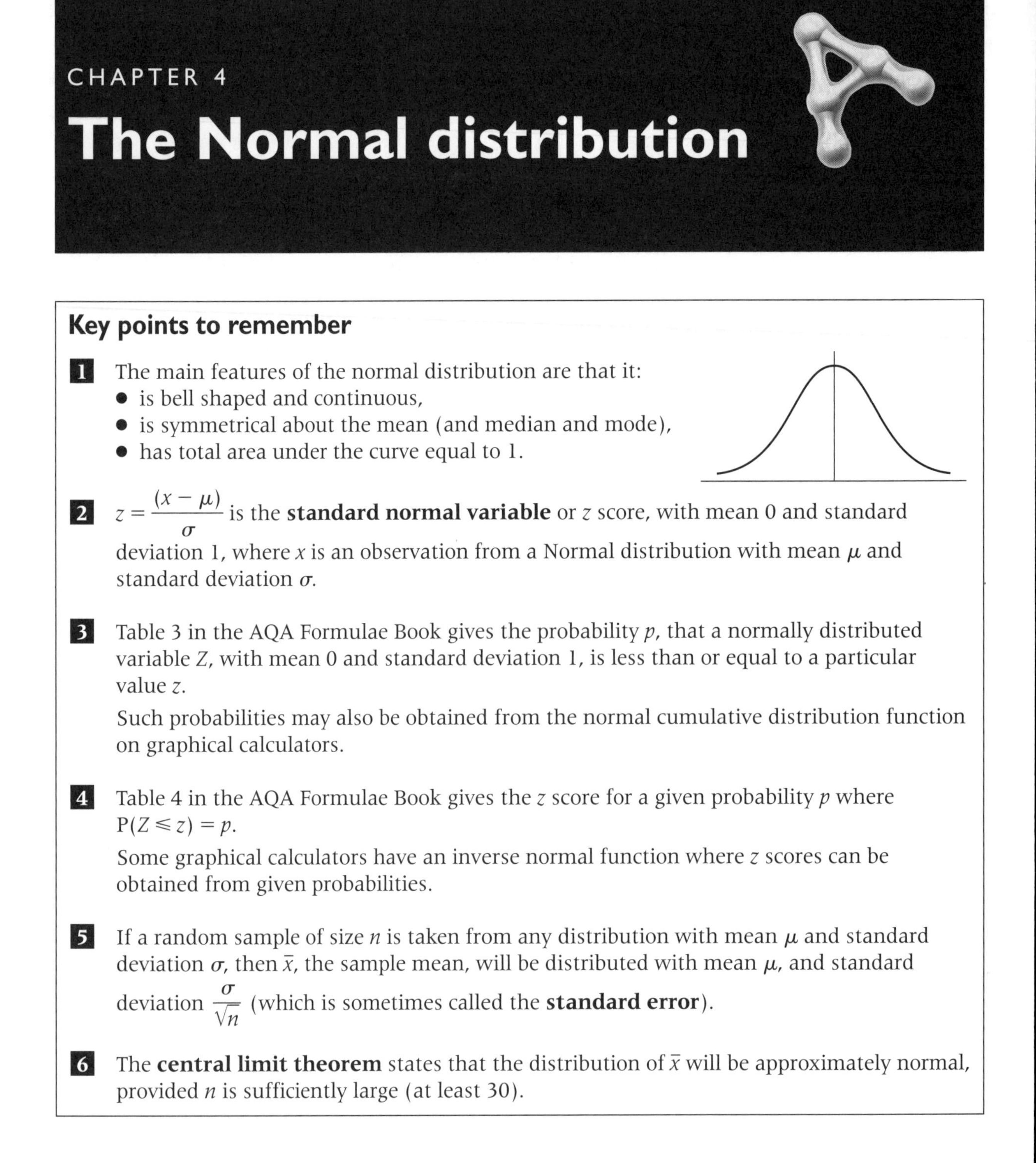

CHAPTER 4

The Normal distribution

Key points to remember

1 The main features of the normal distribution are that it:
- is bell shaped and continuous,
- is symmetrical about the mean (and median and mode),
- has total area under the curve equal to 1.

2 $z = \dfrac{(x - \mu)}{\sigma}$ is the **standard normal variable** or z score, with mean 0 and standard deviation 1, where x is an observation from a Normal distribution with mean μ and standard deviation σ.

3 Table 3 in the AQA Formulae Book gives the probability p, that a normally distributed variable Z, with mean 0 and standard deviation 1, is less than or equal to a particular value z.

Such probabilities may also be obtained from the normal cumulative distribution function on graphical calculators.

4 Table 4 in the AQA Formulae Book gives the z score for a given probability p where $P(Z \leqslant z) = p$.

Some graphical calculators have an inverse normal function where z scores can be obtained from given probabilities.

5 If a random sample of size n is taken from any distribution with mean μ and standard deviation σ, then $\bar{x}$, the sample mean, will be distributed with mean μ, and standard deviation $\dfrac{\sigma}{\sqrt{n}}$ (which is sometimes called the **standard error**).

6 The **central limit theorem** states that the distribution of $\bar{x}$ will be approximately normal, provided n is sufficiently large (at least 30).

Worked example 1

Find the probability that an observation, z, from a standard normal distribution will be:

(a) less than 1.25,
(b) at most 1.156,
(c) less than -1.14,
(d) at most -0.854,
(e) more than 0.96,
(f) more than -1.56,
(g) between 0.96 and 1.15,
(h) between -1.14 and 1.25.

(a) $z = 1.25$ gives the answer $p = 0.894\,35$.
In the exam, the final answer should be quoted to 3 s.f., so $p = 0.894$.

Using **3**

(b) There is a problem using Table 3 here because values for z are only available for z values to a maximum of 2 d.p. In this case, $z = 1.16$ will be used and $p = 0.876\,98$.
From the calculator, using $z = 1.156$, the value of $p = 0.876\,16$. Note that, to 3 s.f., the answers obtained for p differ with Table 3 giving $p = 0.877$ and the calculator giving $p = 0.876$. However, either answer would be acceptable in the exam.

Using **3**

(c) In Table 3, -1.14 is not available as a value for z. However, due to the symmetry of the Normal distribution, $P(Z < -1.14) = P(Z > 1.14)$, and this is obtained by finding
$1 - P(Z < 1.14) = 1 - 0.872\,86 = 0.127$.

Using **1**

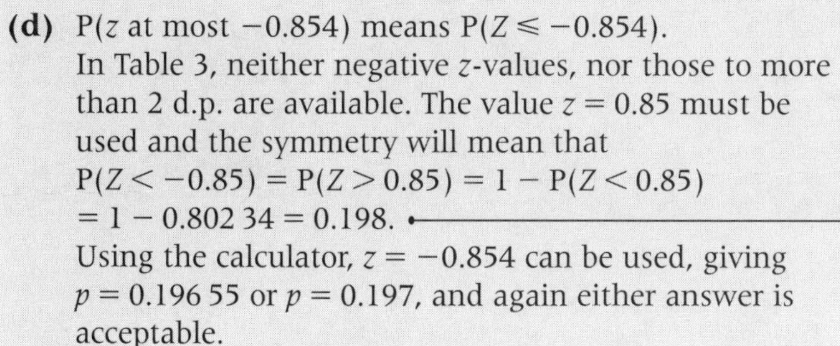

(d) $P(z \text{ at most } -0.854)$ means $P(Z \leqslant -0.854)$.
In Table 3, neither negative z-values, nor those to more than 2 d.p. are available. The value $z = 0.85$ must be used and the symmetry will mean that
$P(Z < -0.85) = P(Z > 0.85) = 1 - P(Z < 0.85)$
$= 1 - 0.802\,34 = 0.198$.
Using the calculator, $z = -0.854$ can be used, giving $p = 0.196\,55$ or $p = 0.197$, and again either answer is acceptable.

Using **3**

(e) First obtain $P(Z \leqslant 0.96)$, and then $P(Z > 0.96)$
$= 1 - P(Z \leqslant 0.96) = 1 - 0.83147 = 0.169$.

(f) By symmetry $P(Z > -1.56) = P(Z \leqslant 1.56)$
which gives the answer $p = 0.941$.

(g) $P(0.96 \leqslant Z \leqslant 1.15)$ can be obtained by finding
$P(Z \leqslant 1.15) - P(Z \leqslant 0.96) = 0.87493 - 0.83147$
$= 0.0435$.

(h) $P(-1.14 \leqslant Z \leqslant 1.25)$ can be obtained by finding
$P(Z \leqslant 1.25) - P(Z \leqslant -1.14)$
$= P(Z \leqslant 1.25) - [1 - P(Z \leqslant 1.14)]$
$= 0.89435 - [1 - 0.87286]$
$= 0.767$.

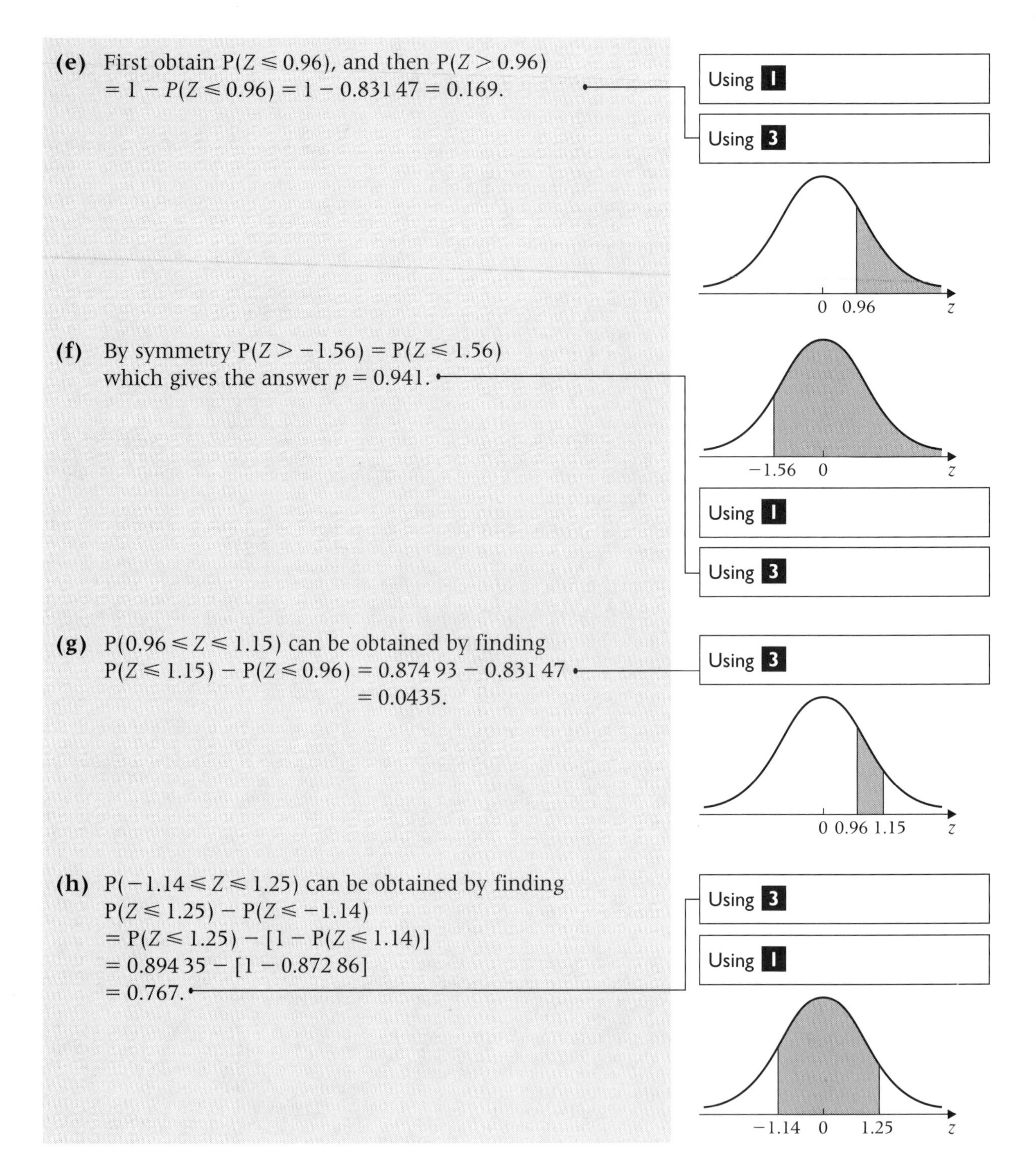

Worked example 2

Find the z score which:

(a) exceeds 90% of the population,

(b) is exceeded by 5% of the population,

(c) is greater than 99% of the population,

(d) is less than 15% of the population,

(e) is exceeded by 30% of the population,

(f) exceeds 20% of the population,

(g) is exceeded by 99.9% of the population.

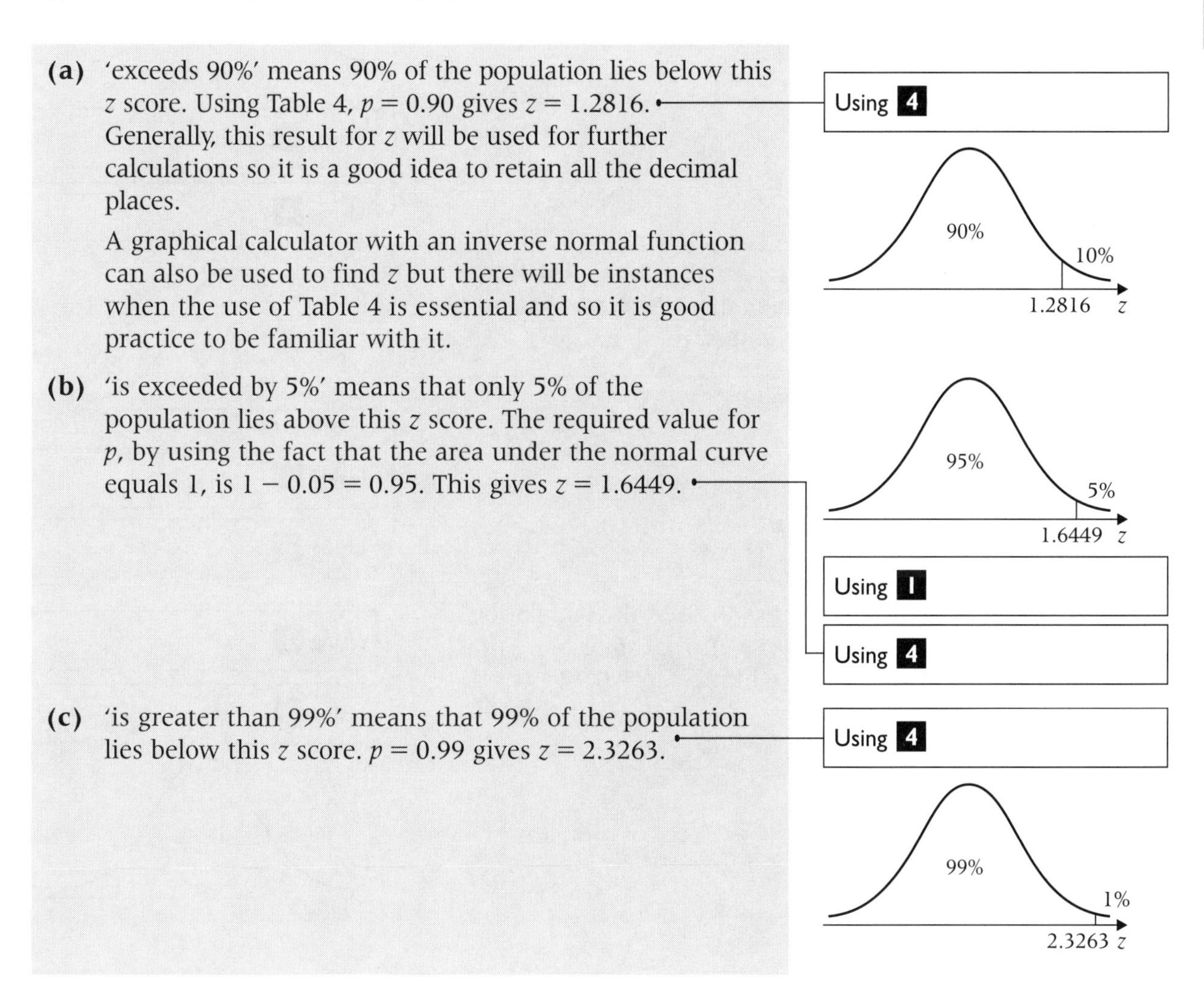

(a) 'exceeds 90%' means 90% of the population lies below this z score. Using Table 4, $p = 0.90$ gives $z = 1.2816$. Generally, this result for z will be used for further calculations so it is a good idea to retain all the decimal places.

A graphical calculator with an inverse normal function can also be used to find z but there will be instances when the use of Table 4 is essential and so it is good practice to be familiar with it.

(b) 'is exceeded by 5%' means that only 5% of the population lies above this z score. The required value for p, by using the fact that the area under the normal curve equals 1, is $1 - 0.05 = 0.95$. This gives $z = 1.6449$.

(c) 'is greater than 99%' means that 99% of the population lies below this z score. $p = 0.99$ gives $z = 2.3263$.

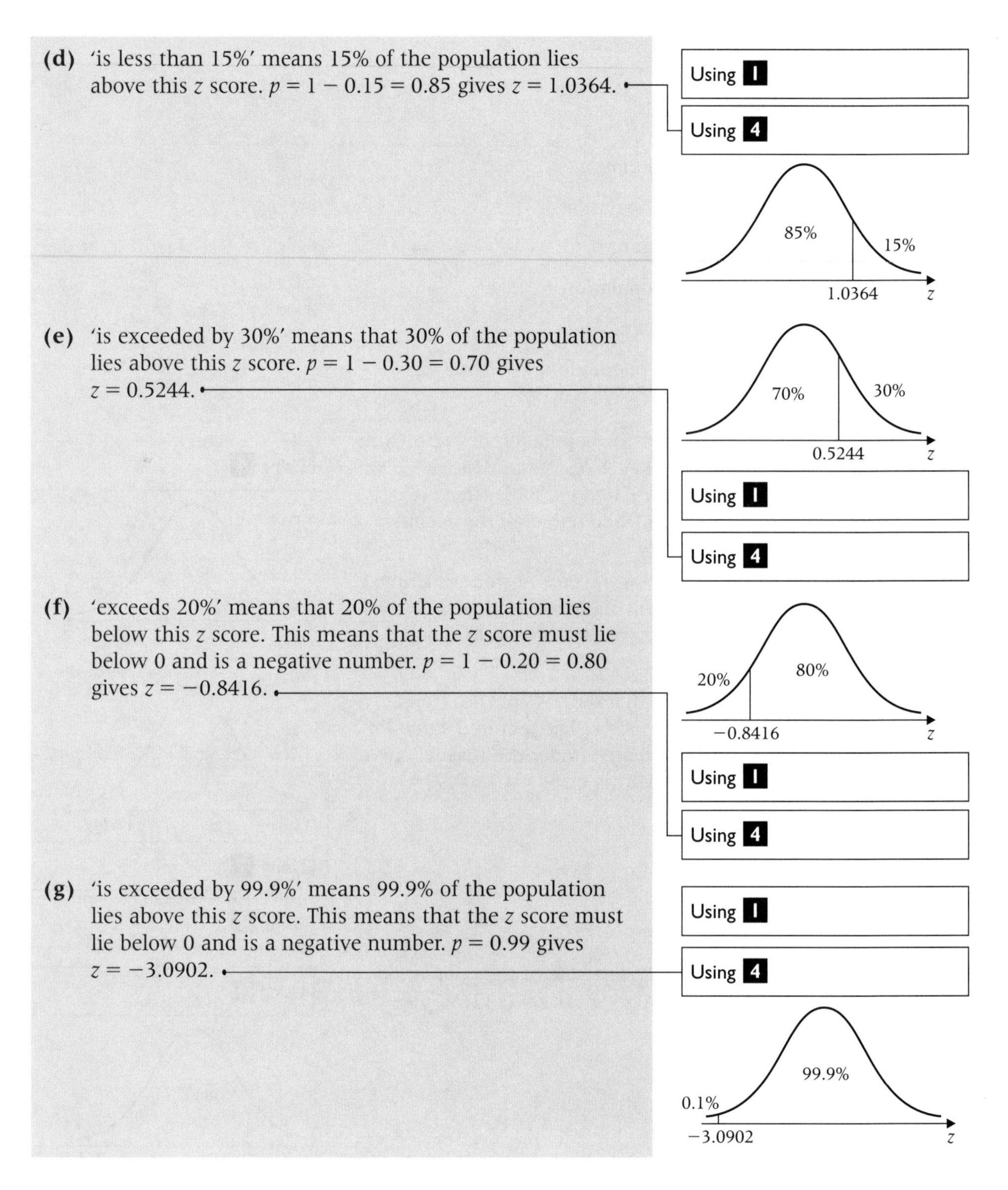

(d) 'is less than 15%' means 15% of the population lies above this z score. $p = 1 - 0.15 = 0.85$ gives $z = 1.0364$.

(e) 'is exceeded by 30%' means that 30% of the population lies above this z score. $p = 1 - 0.30 = 0.70$ gives $z = 0.5244$.

(f) 'exceeds 20%' means that 20% of the population lies below this z score. This means that the z score must lie below 0 and is a negative number. $p = 1 - 0.20 = 0.80$ gives $z = -0.8416$.

(g) 'is exceeded by 99.9%' means 99.9% of the population lies above this z score. This means that the z score must lie below 0 and is a negative number. $p = 0.99$ gives $z = -3.0902$.

Worked example 3

An airline operates a service between London and Glasgow. The flight times, T, can be modelled by a Normal distribution with mean 65 minutes and standard deviation 10 minutes.

(a) Find the probability that a particular flight is:
(i) less than 60 minutes, (ii) no more than 70 minutes,
(iii) at most 72 minutes, (iv) more than 68 minutes,
(v) at least 72 minutes.

(b) Find the flight time T exceeded by:
(i) 90% of these flights, (ii) 10% of these flights,
(iii) 40% of these flights.

(a) We are told that $\mu = 65$ and $\sigma = 10$.
Corresponding z scores must be obtained in order for the values of p to be found, unless a graphical calculator with the facility to relate a value of x, with a known μ and σ, directly to p is used.

Using **2**

(i) $P(T < 60) = P\left(Z < \frac{\{60 - 65\}}{10}\right) = P(Z < -0.5)$

Using **3**

$P(Z < -0.5) = P(Z > 0.5) = 1 - P(Z \leqslant 0.5)$

Using **1**

Which gives $P(T < 60) = 1 - 0.691\,46 = 0.309$.

(ii) $P(\text{no more than } 70) = P(T \leqslant 70) = P\left(Z \leqslant \frac{\{70 - 65\}}{10}\right)$
$= P(Z \leqslant 0.5) = 0.691\,46$ (from (i)) $= 0.691$.

(iii) $P(T \text{ at most } 72) = P(T \leqslant 72) = P\left(Z \leqslant \frac{\{72 - 65\}}{10}\right)$
$= P(Z \leqslant 0.7) = 0.758$

(iv) $P(T > 68) = P\left(Z > \frac{\{68 - 65\}}{10}\right) = P(Z > 0.3)$

Using **1**

$= 1 - P(Z \leqslant 0.3) = 1 - 0.617\,91 = 0.382$

(v) $P(T \text{ at least } 72) = P(T \geqslant 72) = P\left(Z \geqslant \frac{\{72 - 65\}}{10}\right)$
$P(Z \geqslant 0.7) = 1 - P(Z \leqslant 0.7) = 1 - 0.758\,04$
$= 0.242$

(b) For these questions z scores will require conversion into corresponding times.
Some graphical calculators with an inverse normal function can also be used to find T directly.

(i) 90% of flights take longer than this time meaning 90% of z scores lie above this value. This means that the z score must lie below 0 and is a negative number. $p = 0.90$ gives $z = -1.2816$ meaning $\frac{\{T - 65\}}{10} = -1.2816$, and so
$T = 65 - (1.2816 \times 10) = 52.2$ min.

(ii) 10% of flights take longer than this time meaning 10% of z scores lie above this value. $p = 0.90$ gives $z = 1.2816$ meaning $\frac{\{T - 65\}}{10} = 1.2816$, and so
$T = 65 + (1.2816 \times 10) = 77.8$ min.

(iii) 40% of flights take longer than this time meaning 40% of z scores lie above this value. $p = 0.60$ gives $z = 0.2533$ meaning $\frac{\{T - 65\}}{10} = 0.2533$, and so
$T = 65 + (0.2533 \times 10) = 67.5$ min.

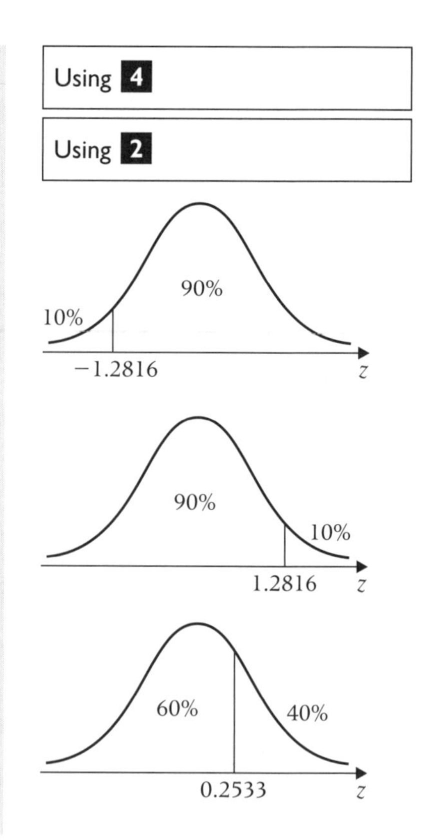

Worked example 4

The time, T minutes, spent in the gym by a member of a club has a mean of 20 and a standard deviation of 5. It can be assumed that T may be modelled by a Normal distribution.

(a) Find the value of T exceeded by 20% of members.

(b) Find the value of T exceeded by 75% of members.

(c) Find the probability that a particular member will spend more than 25 minutes in the gym.

(d) Find the probability that the mean time spent in the gym by a random sample of 10 members will exceed 25 minutes.

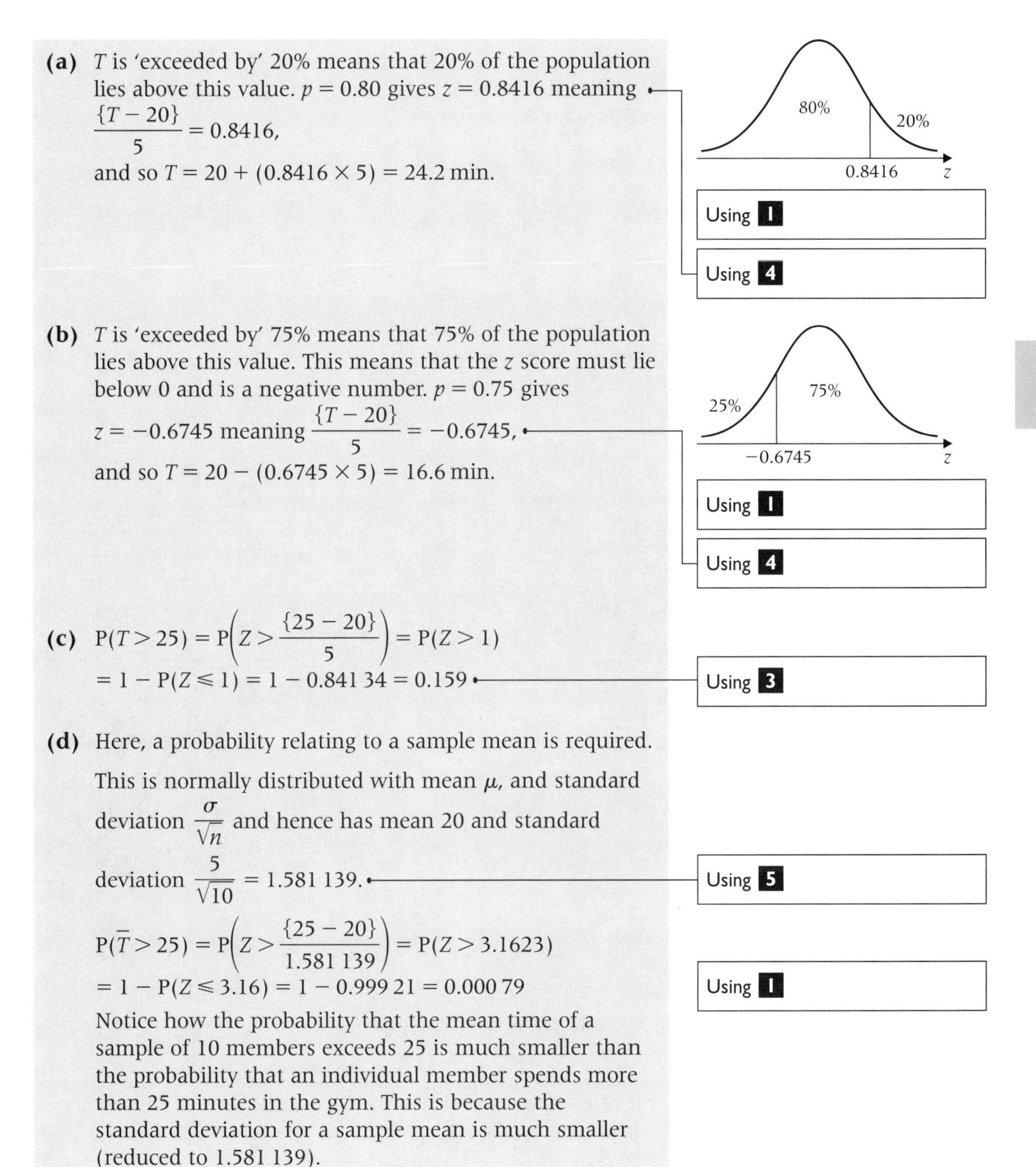

(a) T is 'exceeded by' 20% means that 20% of the population lies above this value. $p = 0.80$ gives $z = 0.8416$ meaning $\frac{\{T - 20\}}{5} = 0.8416$,
and so $T = 20 + (0.8416 \times 5) = 24.2$ min.

(b) T is 'exceeded by' 75% means that 75% of the population lies above this value. This means that the z score must lie below 0 and is a negative number. $p = 0.75$ gives $z = -0.6745$ meaning $\frac{\{T - 20\}}{5} = -0.6745$,
and so $T = 20 - (0.6745 \times 5) = 16.6$ min.

(c) $\mathrm{P}(T > 25) = \mathrm{P}\left(Z > \frac{\{25 - 20\}}{5}\right) = \mathrm{P}(Z > 1)$
$= 1 - \mathrm{P}(Z \leqslant 1) = 1 - 0.841\,34 = 0.159$

(d) Here, a probability relating to a sample mean is required.

This is normally distributed with mean μ, and standard deviation $\frac{\sigma}{\sqrt{n}}$ and hence has mean 20 and standard deviation $\frac{5}{\sqrt{10}} = 1.581\,139$.

$\mathrm{P}(\overline{T} > 25) = \mathrm{P}\left(Z > \frac{\{25 - 20\}}{1.581\,139}\right) = \mathrm{P}(Z > 3.1623)$
$= 1 - \mathrm{P}(Z \leqslant 3.16) = 1 - 0.999\,21 = 0.000\,79$

Notice how the probability that the mean time of a sample of 10 members exceeds 25 is much smaller than the probability that an individual member spends more than 25 minutes in the gym. This is because the standard deviation for a sample mean is much smaller (reduced to 1.581 139).

4

Worked example 5

The lengths, R mm, of rods produced by a manufacturing process are normally distributed with mean 4.2 and standard deviation 0.4.

(a) If the maximum acceptable length for such rods is 4.5 mm, estimate the proportion of rods that are rejected because they are too long.

(b) After an overhaul, it is believed that the machine used to produce these rods is now more reliable. It is found that only 12% of rods are now rejected for being too long. Assuming the mean is still 4.2 mm, find the new standard deviation.

(a) $P(R > 4.5) = P\left(Z > \frac{\{4.5 - 4.2\}}{0.4}\right) = P(Z > 0.75)$

$P(Z > 0.75) = 1 - P(Z \leqslant 0.75)$

Hence $P(R > 4.5) = 1 - 0.773\,37 = 0.227.$

Using **3**

Using **1**

(b) Here, σ is unknown but $\mu = 4.2$ mm.

Now $P(R > 4.5) = 0.12$ (12% of rods are too long)

$P(Z > z) = 0.12$ implies that $P(Z \leqslant z) = 0.88$

which gives $z = 1.1750.$

Hence, $\frac{4.5 - 4.2}{\sigma} = 1.1750$ or $0.3 = 1.1750\sigma$ and

$\sigma = 0.255$ mm.

Using **1**

Using **4**

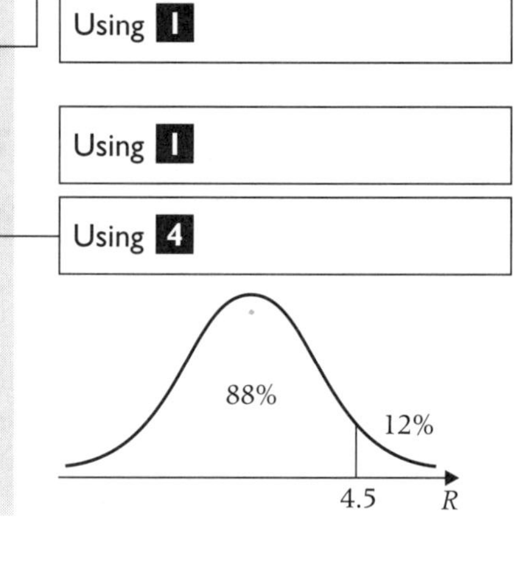

Worked example 6

A water company employs engineers who deal with supply pipe emergencies. The time, T minutes, taken to repair a supply pipe emergency has a mean of 85 and a standard deviation of 75.

(a) Find the probability that the mean time taken to deal with a random sample of 70 supply pipe emergencies is less than 90 minutes.

(b) (i) Explain why it is unlikely that the normal distribution would provide an adequate model for the times taken to deal with supply pipe emergencies.

(ii) Give a reason why it was not necessary to assume that the times are normally distributed in order to find the answer to part **(a)**.

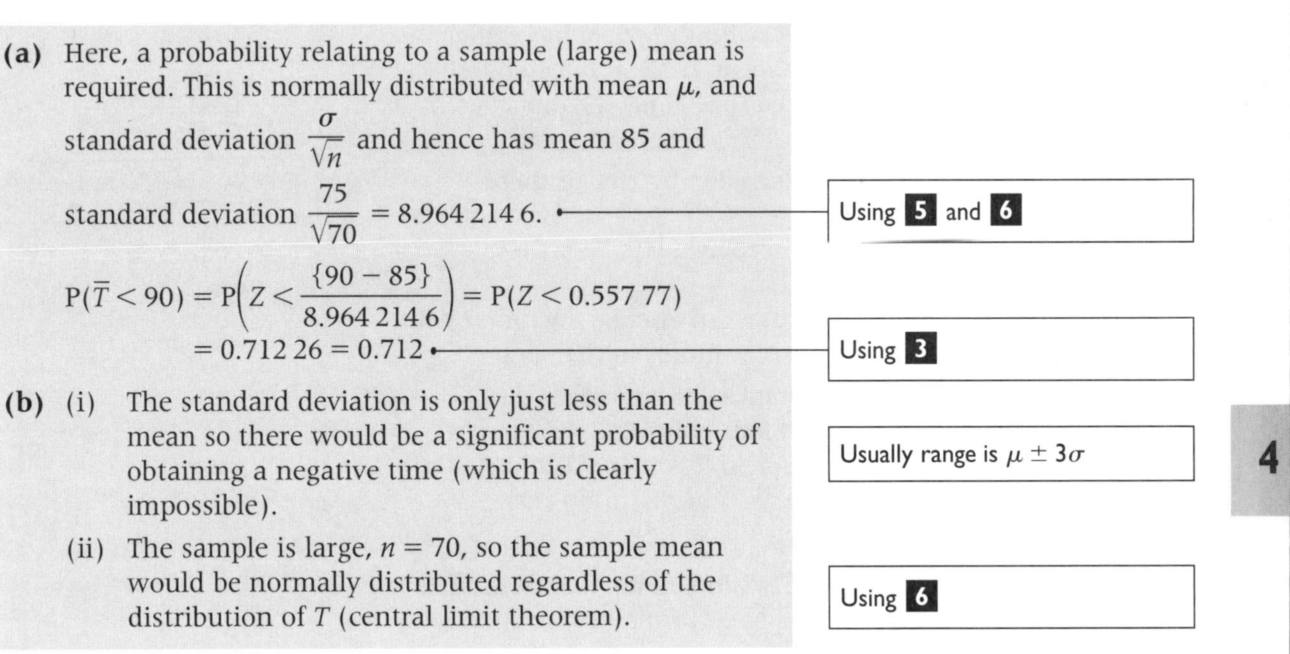

(a) Here, a probability relating to a sample (large) mean is required. This is normally distributed with mean μ, and standard deviation $\frac{\sigma}{\sqrt{n}}$ and hence has mean 85 and standard deviation $\frac{75}{\sqrt{70}} = 8.964\,214\,6$.

Using **5** and **6**

$$P(\overline{T} < 90) = P\left(Z < \frac{\{90 - 85\}}{8.964\,214\,6}\right) = P(Z < 0.557\,77)$$
$$= 0.712\,26 = 0.712$$

Using **3**

(b) (i) The standard deviation is only just less than the mean so there would be a significant probability of obtaining a negative time (which is clearly impossible).

Usually range is $\mu \pm 3\sigma$

(ii) The sample is large, $n = 70$, so the sample mean would be normally distributed regardless of the distribution of T (central limit theorem).

Using **6**

REVISION EXERCISE 4

1 The heights, H, of males attending a school can be modelled by a Normal distribution with mean 171 cm and standard deviation 5.5 cm.

(a) Find the probability that the height of a randomly selected male:

(i) exceeds 155 cm,

(ii) exceeds 175 cm,

(iii) is less than 180 cm,

(iv) is more than 170 cm.

(b) Find the height, H, exceeded by:

(i) 20% of male students,

(ii) 90% of male students.

2 The weights, W g, of lettuces sold at a supermarket can be modelled by a normal distribution with mean 550 and standard deviation 18.

For a randomly selected lettuce, find the probability that its weight:

(a) exceeds 525 g,

(b) is below 510 g,

(c) lies between 530 and 570 g.

(d) Find the weight exceeded by 7% of the lettuces.

3 The time, T minutes, spent in a swimming pool by regular early morning swimmers has a mean of 25 and a standard deviation of 6. It can be assumed that T may be modelled by a Normal distribution.

(a) (i) Find the value of T exceeded by 25% of these swimmers.

(ii) Find the value of T exceeded by 95% of these swimmers.

(iii) Find the probability that a particular swimmer will spend no more than 16 minutes in the pool.

(b) The pool is closed for 3 months for upgrading and modernisation. It is found that only 10% of early morning swimmers now spend less than 21 minutes in the pool. Find the new mean time, assuming the standard deviation remains the same.

4 An airline operates a service between London and Nice. The flight times, T, can be modelled by a Normal distribution with mean 125 minutes and standard deviation 15 minutes.

(a) Find the probability that a particular flight takes:

(i) less than 140 minutes,

(ii) at most 120 minutes,

(iii) more than 148 minutes.

(b) Following a time review, the standard deviation of times for this flight is reduced so that only 5% of flights take longer than 145 minutes. Find the new standard deviation, assuming that the mean remains the same.

5 The weights, W g, of grapefruit sold at a supermarket can be modelled by a Normal distribution with mean 230 and standard deviation 15.
For a randomly selected grapefruit, find the probability that its weight:

(a) (i) exceeds 235 g,

(ii) is below 210 g,

(iii) is between 220 and 230 g.

(b) Find the probability that the mean weight of a randomly selected sample of 10 grapefruit is less than 240 g.

(c) A new supplier is found for grapefruit and it is noted that 20% have weights greater than 245 g and only 5% have weights below 210 g. Find the value of the mean and standard deviation for grapefruit from this new supplier.

6 A water cooler delivers cups of water into a cup when a button is pressed. The volume delivered may be modelled by a Normal distribution with mean 260 ml and standard deviation 15 ml.

(a) Find the probability that the water delivered will be:
(i) less than 270 ml,
(ii) between 250 ml and 270 ml.

(b) Find the probability that the mean volume of water delivered on a random sample of 20 occasions is less than 265 ml.

The cups used to receive the water have a capacity of 275 ml.

(c) Find the probability that when the button is pressed, the volume of water delivered will exceed the capacity of the cup.

The water cooler can be adjusted so that the mean amount of water delivered can take any required value but the standard deviation remains unchanged.

(d) Find the mean value for the volume of water delivered so that the probability of a cup overflowing is 0.01.

7 In order not to be late for work, Amy must leave her house in a taxi no later than 8.30 am. Past experience has shown that, when she telephones for a taxi, the time it takes to arrive may be modelled by a Normal distribution with mean 10 minutes and standard deviation 4 minutes.

(a) Given that Amy telephones for a taxi at 8.15 am, find the probability that she will not be late for work.

(b) Find, to the nearest minute, the latest time that Amy should telephone for a taxi in order to have a probability of 0.99 of not being late for work.

8 Ma Xin uses her mobile phone for T minutes every day. T is a random variable which may be modelled by a Normal distribution with mean 42 and standard deviation 12.

(a) Find the probability that, on a particular day, Ma Xin uses her mobile phone for:
(i) less than 35 minutes,
(ii) between 25 and 35 minutes.

(b) Calculate an interval, symmetrical about the mean, within which T will lie on 90% of days.

(c) Find the probability that, on 14 randomly selected days, the mean time spent by Ma Xin on her mobile phone is at least 40 minutes.

Test yourself

Test yourself	What to review
	If your answer is incorrect:
1 For a standard Normal distribution, find the value of z which is exceeded by: **(a)** 0.05, **(b)** 0.95, **(c)** 0.75, **(d)** 0.15 of the population.	see p 29 Example 2 or review Advancing Maths for AQA S1 pp 89–91
2 Medium eggs have a mean weight of 50 g and a standard deviation of 4 g. Assuming the weights of medium eggs can be modelled by a Normal distribution, find the probability that: **(a)** a randomly selected egg will weight at most 55 g, **(b)** the mean weight of six randomly selected eggs will be more than 55 g.	See p 32 Example 4 or review Advancing Maths for AQA S1 pp 95–98
3 The lengths of rods, L cm, produced in a manufacturing process have mean 4.5 and standard deviation 0.4. Find the probability that a randomly selected rod has length: **(a)** exceeding 4.45 cm, **(b)** no more than 4.42 cm, **(c)** between 4.47 and 4.52 cm.	See p 31 Example 3 or review Advancing Maths for AQA S1 p 87
4 A random sample of 50 packets of crisps is obtained. It is known that the weight of crisps in such packets has mean 37.5 g and standard deviation 2.5 g. **(a)** Find the probability that the mean weight of this sample of 50 packets of crisps exceeds 38 g. **(b)** How would your answer to part **(a)** be affected if it was known that the Normal distribution was not a good model for the weights of crisps. **(c)** How would your answer to part **(a)** be affected if it was known that the sample was not selected at random.	See p 34 Example 6 or review Advancing Maths for AQA S1 pp 95–99

Test yourself ANSWERS

1 **(a)** 1.6449 **(b)** −1.6449 **(c)** −0.6745 **(d)** 1.0364

2 **(a)** 0.894 **(b)** 0.001 10

3 **(a)** 0.550 **(b)** 0.421 **(c)** 0.0498

4 **(a)** 0.0787

(b) No effect as sample size large so central limit theorem states sample mean is normally distributed.

(c) Assumption of normality cannot be made so answer not valid.

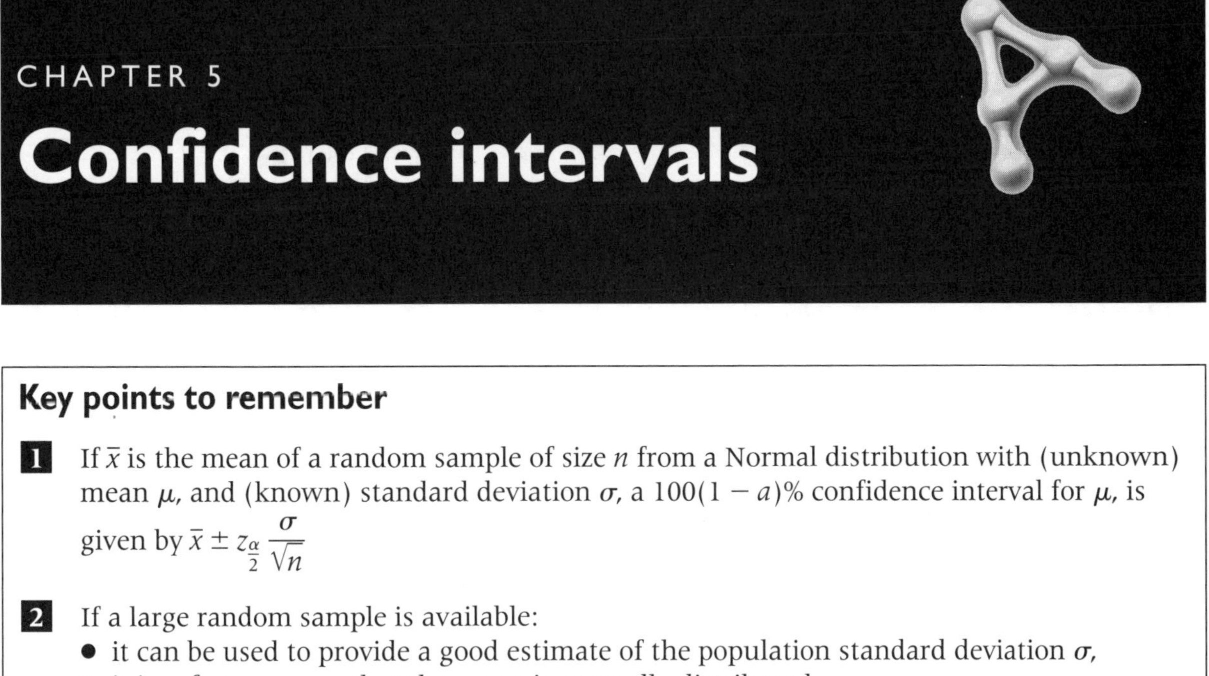

CHAPTER 5

Confidence intervals

Key points to remember

1 If $\bar{x}$ is the mean of a random sample of size n from a Normal distribution with (unknown) mean μ, and (known) standard deviation σ, a $100(1 - a)\%$ confidence interval for μ, is given by $\bar{x} \pm z_{\frac{\alpha}{2}} \dfrac{\sigma}{\sqrt{n}}$

2 If a large random sample is available:
- it can be used to provide a good estimate of the population standard deviation σ,
- it is safe to assume that the mean is normally distributed.

Worked example 1

A large consignment of jars of coffee is delivered to a supermarket. The contents of the jars are known to be normally distributed with standard deviation 1.5 g. Six jars, selected at random, have mean contents of 102.5 g. Calculate a 95% confidence interval for the mean contents of the jars in the consignment.

95% confidence interval for mean is

$$102.5 \pm 1.96 \times \frac{1.5}{\sqrt{6}}$$

i.e. 102.5 ± 1.200 or 101.30 to 103.70.

Using **1**

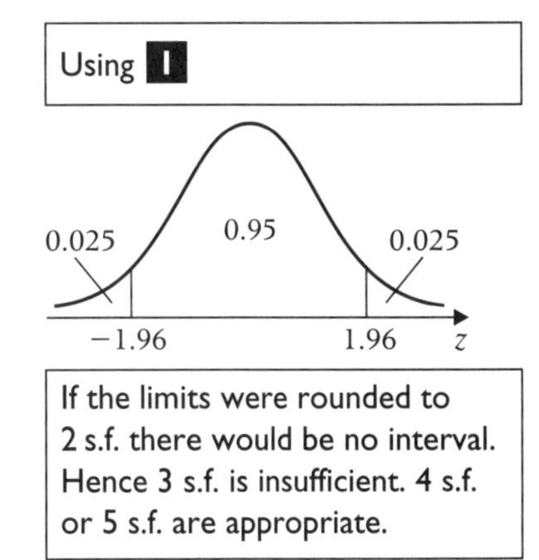

If the limits were rounded to 2 s.f. there would be no interval. Hence 3 s.f. is insufficient. 4 s.f. or 5 s.f. are appropriate.

Worked example 2

A random sample of 160 car batteries from a particular manufacturer were monitored. The mean lifetime, X, of each of these batteries was recorded.

The results are summarised below.

$\Sigma x = 6608 \quad \Sigma(x - \bar{x})^2 = 4466$ where $\bar{x}$ is the sample mean.

(a) Calculate a 90% confidence interval for the mean lifetime of this manufacturer's batteries.

(b) Explain why it was possible to answer part **(a)** without knowing the distribution of the lifetimes.

(a) $\bar{x} = \frac{6608}{160} = 41.3$

$s^2 = \frac{4466}{159} = 28.088\,05,\ s = 5.2998$

90% confidence interval for the mean lifetime is

$$41.3 \pm 1.6449 \times \frac{5.2998}{\sqrt{160}}$$

i.e. 41.3 ± 0.689 or 40.61 to 41.99.

(b) As the sample is large the central limit theorem states that the sample mean will be approximately normally distributed whatever the distribution of the parent population.

Using **1**

Use the divisor $n - 1$ to estimate the standard deviation.

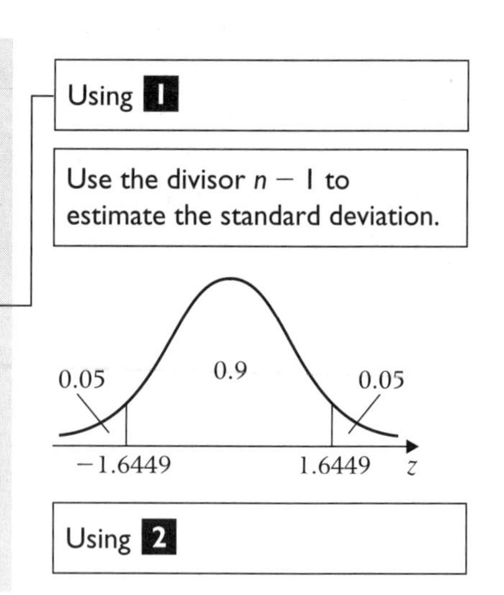

Using **2**

Worked example 3

A hospital kitchen buys oranges in large batches. In each batch, the vitamin C content of the flesh of the oranges, in milligrams per 10 grams, may be modelled by a Normal distribution with standard deviation 0.13.

The vitamin C contents of the flesh of a random sample of 8 oranges from a particular batch were measured, with the following results.

1.32 1.11 1.31 1.22 1.25 1.57 1.42 1.36

(a) Calculate a 99% confidence interval for the mean vitamin C content of the flesh of oranges in this batch.

(b) A nutritionist, who draws up diets for individual patients, assumes that the flesh of oranges has a vitamin C content of at least 1.20. Comment on this assumption as it relates to this batch of oranges.

(a) $\bar{x} = 1.32$

99% confidence interval for the mean vitamin C content of the batch of oranges is

$$1.32 \pm 2.5758 \times \frac{0.13}{\sqrt{8}}$$

i.e. 1.32 ± 0.118 or 1.202 to 1.438.

(b) The confidence interval shows that it is safe to assume that the mean vitamin C content of the oranges in the batch will exceed 1.20. However some individual oranges (including one in the sample) will have a vitamin C content of less than 1.20.

Find $\bar{x}$ directly from your calculator.

Using **1**

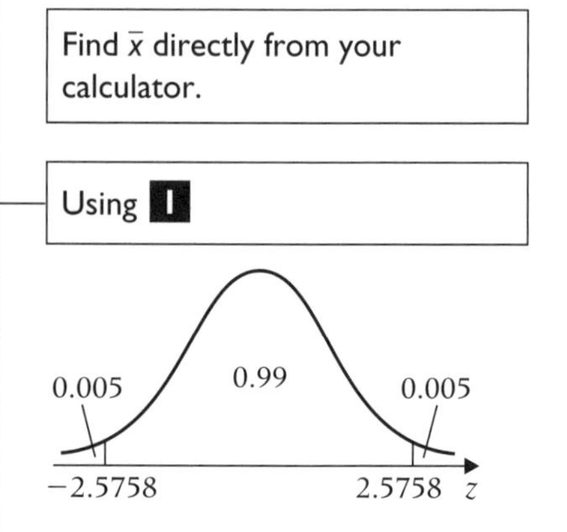

Worked example 4

A company manufactures components for the motor industry. The components are designed to have a length of 135.0 mm. Each day a technician takes a random sample of components and calculates a 95% confidence interval for the mean length. On a particular Monday the technician takes a sample of nine components. Their lengths, in millimetres, are as follows.

135.1 135.7 134.9 135.2 136.3 135.7 135.9 136.0 135.6

(a) Assuming the lengths of components produced on this Monday may be modelled by a Normal distribution with standard deviation 0.42, calculate a 95% confidence interval for the mean length. Give your answer to an appropriate degree of accuracy.

(b) Comment on your result as it relates to the design length of 135.0 mm.

(c) State the width of the confidence interval you have calculated.

(d) What percentage would be associated with a confidence interval of width 0.40 mm calculated from the given data?

(e) State the probability that the 95% confidence interval which the technician will calculate next Monday will not contain the mean length of the components produced on that day.

(a) $\bar{x} = 135.6$

95% confidence interval for the mean is

$$135.6 \pm 1.96 \times \frac{0.42}{\sqrt{9}}$$

Using **1**

i.e. 135.6 ± 0.2744 or 135.33 to 135.87.

(b) The confidence interval provides evidence that the mean length of components produced on this Monday is above the design length of 135.0 mm.

(c) $2 \times 0.2744 = 0.549$

(d) $$2z \times \frac{0.42}{\sqrt{9}} = 0.4$$

$z = 1.429$

$1 - 2 \times 0.0765 = 0.847$

i.e. an 84.7% confidence interval would be of width 0.4 mm.

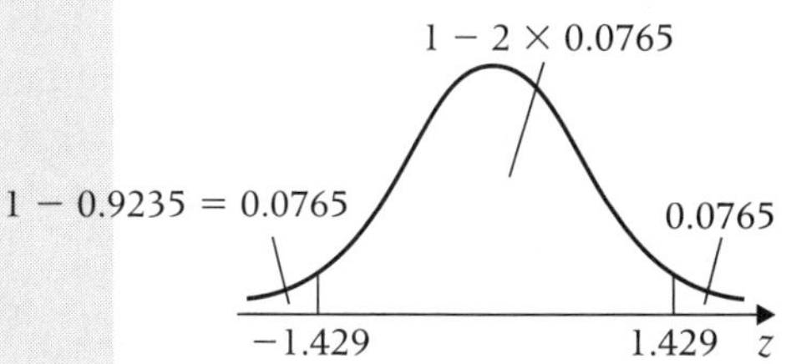

Interpolation has been used to give a more accurate answer. Rounding z to 1.43 is acceptable and would receive full marks.

(e) A 95% confidence interval for the mean is calculated in such a way that 95% of all such intervals will contain the population mean. Hence the probability that a particular confidence interval will not contain the population mean is 0.05.

REVISION EXERCISE 5

1 The weights, in kg, of luggage belonging to seven passengers waiting to board transatlantic flights were as follows.

47 23 62 28 31 75 44

Assume that the data may be regarded as a random sample from a Normal distribution with mean μ and standard deviation 10 kg. Calculate:

(a) an 80% confidence interval for μ,

(b) a 90% confidence interval for μ,

(c) a 95% confidence interval for μ,

(d) a 99% confidence interval for μ,

(e) a 99.8% confidence interval for μ.

2 An electricity company selected a random sample of size 160 from those customers who had not paid their bills five weeks after they had been sent out. The mean amount owed by the customers in the sample was £107.50 and the standard deviation was £31.00.

Calculate a 95% confidence interval for the mean amount owed by all customers who had not paid their bills five weeks after they had been sent out.

3 The sauce in each of a random sample of 112 jars from a large consignment is weighed. These weights are found to have a mean of 98.3 grams and a standard deviation of 1.5 grams.

(a) Calculate a 90% confidence interval for the mean content of jars in the consignment.

(b) Each of the jars has a nominal content of 100 grams. Is the suspicion that the consignment is underweight valid? Justify your answer.

4 A random sample of experimental components for use in four-wheel drive vehicles taking part in the school run in London suburbs were tested to destruction under extreme conditions. The survival times, X days, of ten components were as follows.

217 362 119 645 243 319 990 198 433 550

(a) Assuming that the survival time, under these conditions, for all the experimental components is normally distributed with standard deviation 217 days, calculate a 90% confidence interval for the mean of X.

(b) State the probability that a 90% confidence interval calculated from an independent random sample of survival times will not contain the mean of X.

5 A large shipment of hamburger meat is made up of 1 kg packages. A random sample of 95 of these packages was analysed in a laboratory and the fat content determined. The mean fat content was found to be 670 g with a standard deviation of 35 g.

(a) Calculate a 95% confidence interval for the mean fat content of the packages of hamburger meat.

(b) Explain the relevance of the central limit theorem to your calculation in part **(a)**.

(c) Assuming the fat content of the packages is normally distributed with mean 670 g, calculate an interval within which 95% of the fat contents will lie.

(d) Comment on the shipper's claim that the fat content of the packages is less than 680 g.

6 An instrument measures impurity in a chemical in mg per 100 g of the chemical. Eighty readings, X, are taken on a batch of chemical and the results are summarised below.

$\Sigma x = 736.8 \qquad \Sigma(x - \bar{x})^2 = 814.02$

(a) Calculate estimates for the mean and standard deviation of the impurity readings on this batch of chemical.

(b) Hence calculate a 99% confidence interval for the mean impurity reading on this batch of chemical.

(c) Write down the width of the interval you have calculated in part **(b)**.

(d) What percentage would be associated with an interval calculated from the given data of width

(i) 1.5, (ii) 1.0.

7 The television watching habits of 12-year-old children in a European country are to be studied. The total number of minutes, X, spent watching TV by each of a random sample of 110 children during a particular week was recorded. The results are summarised below.

$\Sigma x = 79\,101 \quad S_{xx} = 6\,278\,642$

(a) Calculate estimates for the mean and standard deviation of X.

(b) Calculate a 90% confidence interval for the mean of X.

(c) Comment on the education minister's claim that on average 12-year-old children watch less than 10 hours TV a week.

(d) How would your answer be affected if it later emerged that:

(i) the sample was random but the distribution of X was not normal,

(ii) the distribution of X was normal but the sample was not random.

Test yourself	What to review
	If your answer is incorrect:
1 23.5 32.4 53.8 19.4 44.3 The above numbers are a random sample of five observations from a Normal distribution with mean μ and standard deviation 7.4. Calculate a 95% confidence interval for μ.	See p 39 Example 1 or review Advancing Maths for AQA S1 p 109
2 A random sample of 120 customers of Greedco Supermarket queued at a checkout for a mean of 235 seconds before being served. The standard deviation of the queuing times was 34 seconds. **(a)** Calculate a 90% confidence interval for the mean queuing time for all Greedco's customers. **(b)** Explain why it was not necessary to assume that the distribution was normal in order to answer part **(a)**.	See p 39 Example 2 or review Advancing Maths for AQA S1 pp 96, 112
3 Following the reorganisation of an ambulance service, the time taken to respond to a random sample of 98 emergency calls had a mean of 8.6 minutes with a standard deviation of 1.9 minutes. **(a)** Calculate a 99% confidence interval for the mean time for ambulances to respond to emergency calls. **(b)** Comment on the ambulance services claim that emergency calls will be answered in less than 8 minutes.	See p 40 Example 3 or review Advancing Maths for AQA S1 pp 109–110 and 112
4 In order to estimate the mean weight of 9-year-old school children in an inner city area, a random sample of children are to be weighed and a 95% confidence interval calculated. **(a)** State the probability that this 95% confidence interval will contain the population mean. **(b)** If a 75% confidence interval for the mean is calculated from the same data, state the probability that this confidence interval will not contain the population mean. **(c)** State the probability that: (i) the 95% confidence interval will contain the population mean but the 75% confidence interval will not, (ii) the 75% confidence interval will contain the population mean but the 95% confidence interval will not.	See p 41 Example 4 or review Advancing Maths for AQA S1 p 117

Test yourself ANSWERS

1 28.2–41.2

2 **(a)** 229.9–240.1

(b) As sample is large the central limit theorem states that the sample mean will be approximately normally distributed whatever the parent distribution.

3 **(a)** 8.11–9.09

(b) The confidence interval provides evidence that the mean time to answer emergency calls is greater than 8 minutes (although some individual emergency calls will be answered in less than 8 minutes).

4 **(a)** 0.95 **(b)** 0.25 **(c)** (i) 0.2 (ii) 0

5

CHAPTER 6

Correlation

Key points to remember

1 A scatter diagram should be drawn to judge whether correlation is present.

2 The **product moment correlation coefficient**,

$$r = \frac{\Sigma(x - \bar{x})(y - \bar{y})}{\sqrt{\Sigma(x - \bar{x})^2(y - \bar{y})^2}}$$

or

$$\frac{S_{xy}}{\sqrt{S_{xx}S_{yy}}}$$

Remember this can be found directly using a calculator.

r is a measure of **linear** relationship only and $-1 \leqslant r \leqslant 1$.
Do not refer to r if a scatter diagram clearly shows a non-linear connection.

3 $r = +1$ or $r = -1$ implies that the points all exactly lie on a straight line.
$r = 0$ implies no linear relationship is present.
But ... no linear relationship between the variables does not necessarily mean that $r = 0$.

4 Even if r is close to $+1$ or -1, no causal link should be assumed between the variables without thinking very carefully about the nature of the data involved.

Worked example 1

The following data is extracted from an investigation into changes in vegetation cover brought about by human activity. The vegetation cover was sampled on a path used by hang-glider pilots climbing to the top of Dunstable Downs. The table records the number of species, x, and the percentage of vegetation cover, y, in nine half-metre squares.

x	4	8	5	7	5	7	9	6	4
y	60	55	30	70	75	35	50	75	45

(a) Draw a scatter diagram of the data.

(b) Calculate the value of the product moment correlation coefficient between x and y.

(c) The investigator expected the number of species to increase with the vegetation cover. Explain to what extent, if at all, your scatter diagram and calculation support the investigator's expectation.

(a)

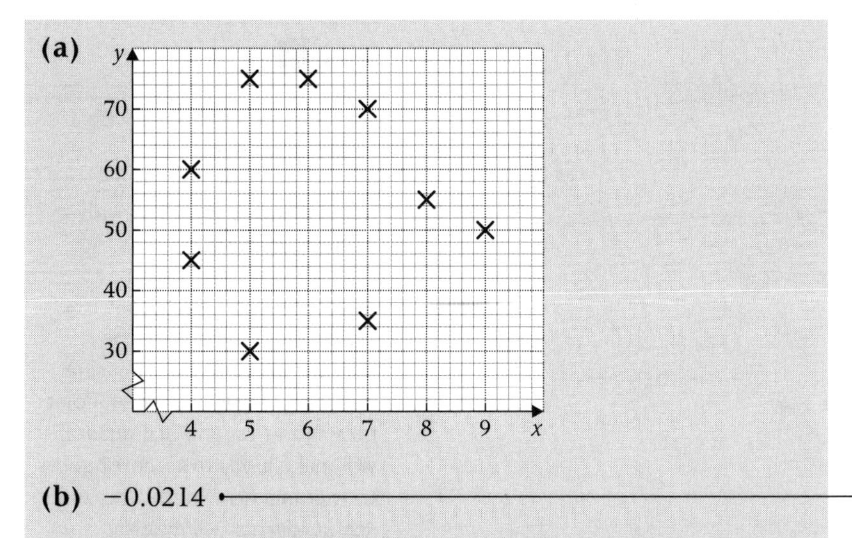

The diagram does not suggest any linear association.

(b) −0.0214

Using **2**

Find r directly from your calculator.

(c) The scatter diagram does not indicate any association between number of species and the vegetation cover. This is confirmed by the magnitude of the correlation coefficient which is very small.

If the magnitude had not been so small, the fact that the correlation coefficient is negative would have suggested that, contrary to the investigator's belief, the number of species reduced as vegetation cover increased.

Using **3**

Worked example 2

Estimate, without undertaking any calculations, the values of the product moment correlation coefficient between the variables X and Y in each of the scatter diagrams below.

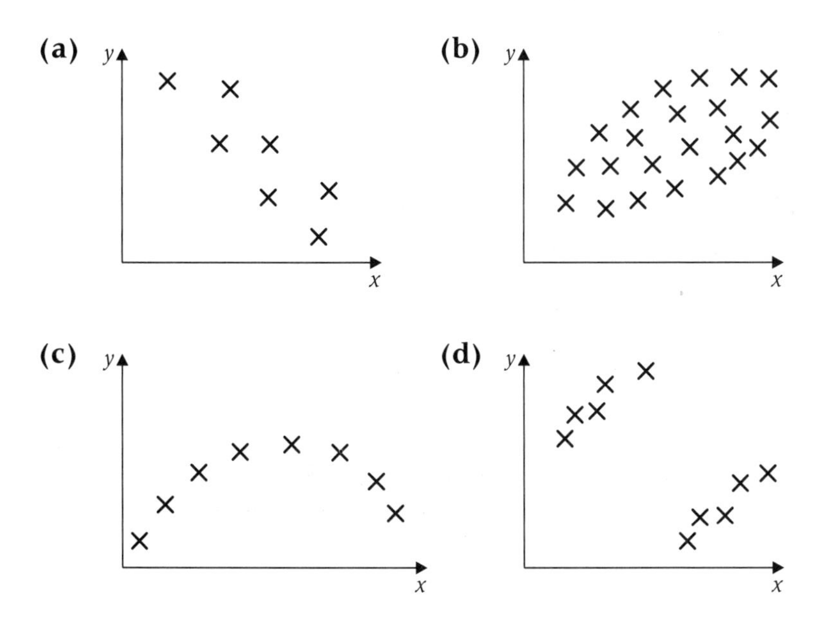

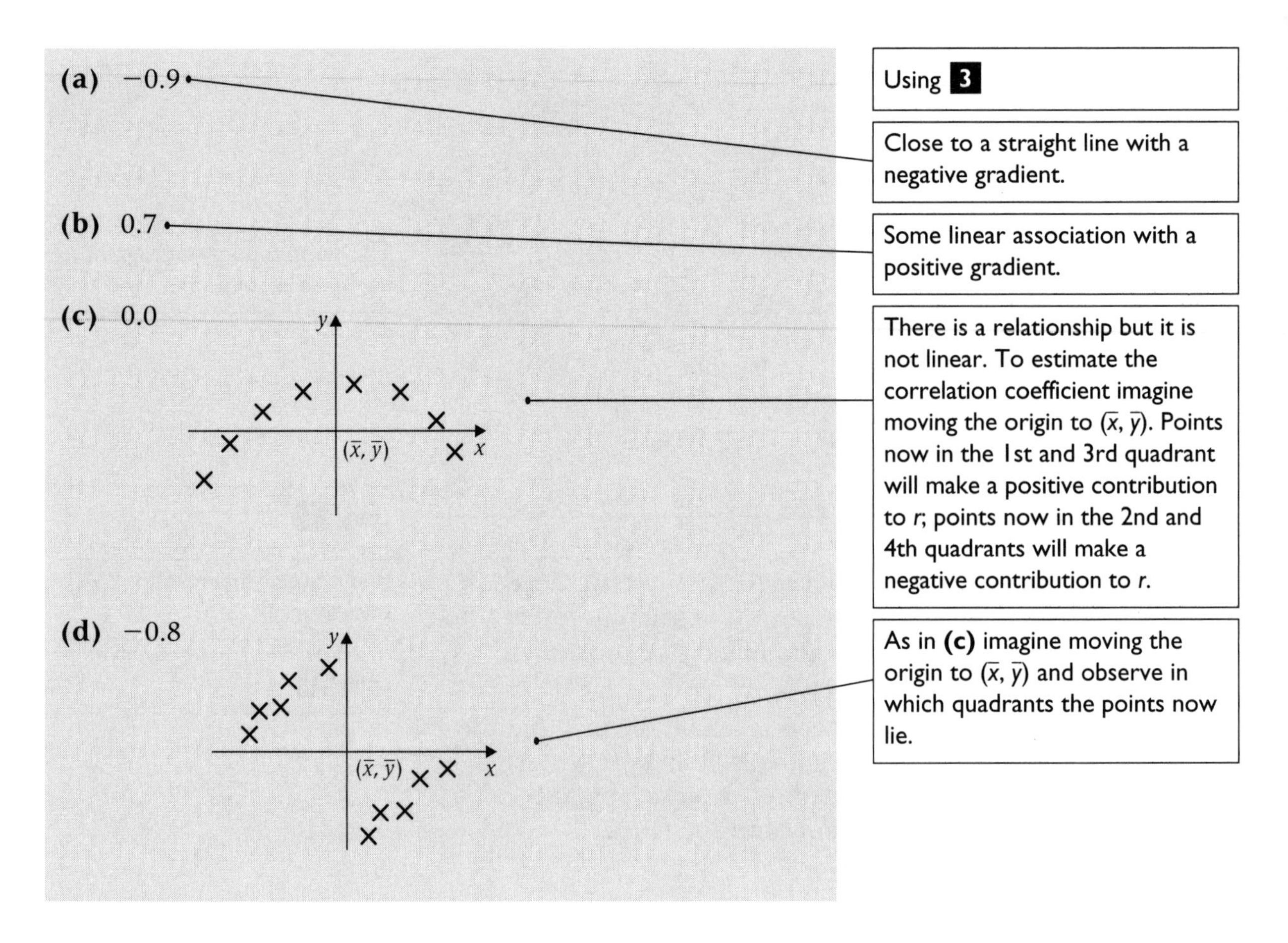

Worked example 3

A class of Statistics students studying correlation obtain data by completing a questionnaire and by referring to the register. They use the data collected to practise their calculations.

A student calculates that the product moment correlation coefficient for the class is:

(a) 1.17 between average journey time to school and distance travelled,

(b) −0.34 between height and weight,

(c) 0.75 between marks in the last test and number of days absent last term.

For each calculation, state, giving a reason, whether the result is probably correct, probably incorrect or definitely incorrect.

(a) Definitely incorrect as the correlation coefficient can only take values in the range −1 to +1.

Using **2**

(b) Probably incorrect as you would expect tall students to be heavy, which would give a positive correlation coefficient.

(c) Probably incorrect as you would expect students with a large number of absences to score low marks, which would give a negative correlation coefficient.

Worked example 4

A cricket team meets for fielding practice. One exercise consists of a cricket ball being thrown at different heights, speeds and angles to one side of a fielder who tries to catch it one-handed.

Each member of the team attempts 25 catches with each hand. The numbers of successful catches are given in the following table.

Fielder	*A*	*B*	*C*	*D*	*E*	*G*	*H*	*I*	*J*	*K*	*L*
Left hand	11	13	9	17	21	16	14	8	19	19	20
Right hand	18	17	20	22	14	19	21	15	10	24	23

(a) Calculate the value of the product moment correlation coefficient between the number of catches with the left hand and the number of catches with the right hand.

(b) Comment on the performance of fielders *E* and *J*.

(c) If fielders *E* and *J* are omitted from the calculation, the value of the product moment correlation coefficient between the number of left handed and the number of right handed catches is 0.812, correct to 3 d.p. Comment on this value and the value you calculated in part **(a)**.

(a) 0.0477

Using **2**

Find r directly from your calculator.

(b) Unlike all other fielders, both *E* and *J* made more catches with their left hand than with their right hand. (Perhaps they were left handed and the rest were right handed.)

(c) A correlation coefficient of 0.812 is high and suggests that those who caught a lot of catches with their right hand also caught a lot of catches with their left hand. A correlation coefficient of 0.0476 is low and does not suggest any association between the number of catches with the left and with the right hand.

6

REVISION EXERCISE 6

1 Calculate the correlation coefficient between the variables x and y in the following cases.

(a)

x	5	4	9	12	17	19
y	3	11	9	10	16	23

(b)

x	17.2	19.6	21.3	14.0	21.3	32.2	29.1	17.2
y	38.6	11.2	9.4	18.5	2.8	1.7	9.0	16.4

(c) $S_{xx} = 292$, $S_{yy} = 374$, $S_{xy} = 187$

(d) $S_{xx} = 99.33$, $S_{yy} = 8.325$, $S_{xy} = -24.32$

(e) $\Sigma(x - \bar{x})^2 = 211$, $\Sigma(y - \bar{y})^2 = 936$, $\Sigma(x - \bar{x})(y - \bar{y}) = -300$

(f) $\Sigma(x - \bar{x})^2 = 0.034\,51$, $\Sigma(y - \bar{y})^2 = 34.22$.
$\Sigma(x - \bar{x})(y - \bar{y}) = 0.777$

2 Estimate, without undertaking any calculations, the values of the product moment correlation coefficient between the variables X and Y in each of the scatter diagrams below.

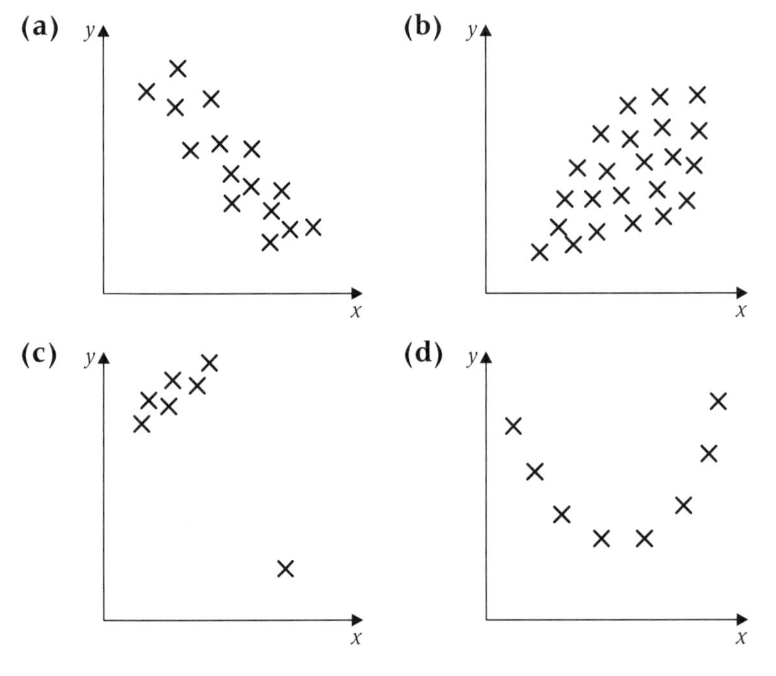

3 The following table shows, for a sample of towns in Great Britain, the number of solicitors, x, and the number of cars stolen last week, y.

x	12	7	11	19	5	21	3	4	17
y	14	3	21	28	6	43	1	12	30

(a) (i) Calculate the value of the product moment correlation coefficient for the data.

(ii) Interpret your result from part **(a)**(i) in the context of the question.

(b) Comment on the suggestion that most car thieves are solicitors.

4 Henri and Michelle are two journalists who write regular newspaper columns advising readers which wines offer good value for money. They taste a number of wines and then estimate the retail prices of the corresponding bottles of wine. Their estimates are shown in the following table.

Henri's estimate (£)	7	3	25	50	2	5	17	80	12
Michelle's estimate (£)	9	19	20	25	9	10	4	26	60

(a) Calculate the value of the product moment correlation coefficient between Henri's and Michelle's estimates.

(b) Interpret, briefly, your value of the correlation coefficient.

5 The following three statements are the results of Damien's calculations for a sample of towns in England.

Statement 1

The product moment correlation coefficient, between the percentage unemployed and the percentage of the population with qualifications beyond GCSE, is 0.65.

Statement 2

The product moment correlation coefficient, between the number of reported crimes of violence in 2003 and the number of chartered accountants, is 1.15.

Statement 3

The product moment correlation coefficient, between the number of reported crimes of violence in 2003 and the number of mathematics teachers, is 0.82.

Classify each of the three statements as either:

A plausible,
B probably incorrect,
C definitely incorrect.

Give a reason for each of your answers.

6 An instrument panel is being designed to control a complex industrial process. It will be necessary to use both hands independently to operate the panel. To help with the design it was decided to time a number of operators, each carrying out the same task once with the left hand and once with the right hand.

The times, in seconds, were as follows.

Operator	*A*	*B*	*C*	*D*	*E*	*F*	*G*	*H*	*I*	*J*	*K*
Left hand, x	49	58	63	42	27	55	39	33	72	66	50
Right hand, y	34	37	49	27	49	40	66	21	64	42	37

(a) Plot a scatter diagram of the data.

(b) Calculate the product moment correlation coefficient between the two variables and comment on this value.

(c) Further investigation revealed that two of the operators were left handed. State, giving a reason, which you think these were. Omitting their two results, calculate the product moment correlation coefficient and comment on this value.

(d) What can you say about the relationship between the times to carry out the task with left and right hands?

7 A local authority offers all its employees regular health checks. As part of the check several physiological measurements are taken on each person. The results for two of the measurements X_1, X_2 on 9 people are shown below.

X_1	9	31	29	50	54	69	76	91	95
X_2	10	21	30	15	34	44	61	51	64

(a) Draw a scatter diagram of the data.

(b) Calculate the product moment correlation coefficient r.

(c) The scatter diagrams below are for X_3, X_4 measured on 8 people and for X_5, X_6 measured on 10 people. The correlation coefficients calculated from these data are respectively 0.90 and 0.88.

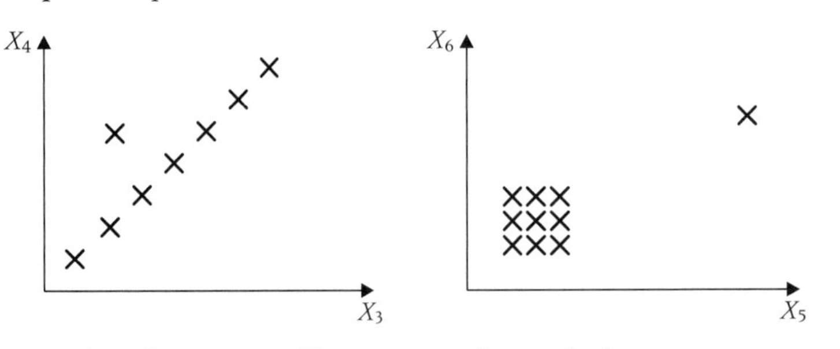

Examine the scatter diagrams and correlation coefficients and compare the three relationships.

Test yourself

What to review

If your answer is incorrect:

1 Calculate the product moment correlation coefficient for the data below.

x	12	32	1	14	45	23
y	7	24	18	3	44	29

See p 46 Example 1 or review Advancing Maths for AQA S1 pp 127–130

2 Calculate the product moment correlation coefficient between s and t.
$\Sigma(s-\bar{s})^2 = 432 \quad \Sigma(t-\bar{t})^2 = 236 \quad \Sigma(s-\bar{s})(t-\bar{t}) = -99$

See p 46 Example 1 or review Advancing Maths for AQA S1 p 128

3 Calculate the product moment correlation coefficient between x and y.
$S_{xx} = 2.41 \quad S_{yy} = 3.56 \quad S_{xy} = 1.87$

See p 46 Example 1 or review Advancing Maths for AQA S1 p 128

4 Plot a scatter diagram of the data below.

x	12	24	32	44	55	67	74
y	4.2	6.3	8.4	12.6	20.8	35.0	54.5

Explain why it would be inappropriate to calculate the product moment correlation coefficient for this data.

See p 47 Example 2 or review Advancing Maths for AQA S1 pp 131–133

5 The following table shows, for a sample of towns, the number of practising dentists, x, and the number of crimes of violence reported in 2004, y.

x	14	27	32	8	28	9	15	41
y	122	276	274	47	209	103	142	366

(a) Calculate the product moment correlation coefficient between x and y.

(b) Does the data support the view that most crimes of violence are committed by dentists? Explain your answer.

See p 46 Example 1 or review Advancing Maths for AQA S1 pp 129–131

Test yourself ANSWERS

1 0.774

2 −0.310

3 0.638

4

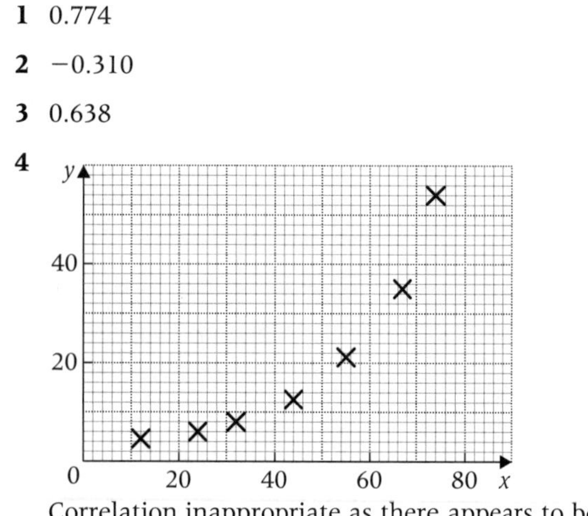

Correlation inappropriate as there appears to be a non-linear relationship between x and y.

5 **(a)** 0.974

(b) Data suggests association between number of dentists and number of crimes of violence. Probably explained by size of town. It does not show that most crimes of violence are committed by dentists.

CHAPTER 7

Regression

Key points to remember

1 A scatter diagram should be drawn to judge whether linear regression analysis is a sensible option.

2 The nature of the data should be considered to determine which is the independent or explanatory variable (x) and which is the dependent or response variable (y).

3 The **regression line** is found using the method of least squares in the form $y = a + bx$. This is the regression line of y on x and may be used to predict a value for y from a given value of x.

The equations can be found directly using a calculator with a linear regression mode. Be careful to note the form in which your calculator presents the equation – it may be as $y = ax + b$.

4 The equation of the regression line is

$$y - \bar{y} = b(x - \bar{x}) \text{ where } b = \frac{\Sigma(x - \bar{x})(y - \bar{y})}{\Sigma(x - \bar{x})^2} \text{ or } \frac{S_{xy}}{S_{xx}}.$$

5 Using $y = a + bx$
a estimates the value of y when x is zero.
b estimates the rate of change of y with x.

6 Be very careful when predicting from your line. Watch out for extrapolation when predictions can be wildly inaccurate. Never assume a linear model will keep on going forever.

7 The residual for each point is the observed value of y minus the value of y predicted by the regression line. For the point (x_i, y_i) the residual $= y_i - a - bx_i$.

7

Worked example 1

Find the equation of the regression line of y on x, given that:

(a) $S_{xx} = 235$, $S_{xy} = 432$, $\bar{x} = 8$, $\bar{y} = 22$,

(b) $\Sigma(x - \bar{x})^2 = 2956$, $\Sigma(x - \bar{x})(y - \bar{y}) = -18.2$ and the line passes through the point $(-7.8, 4.4)$.

(a) $y - 22 = \frac{432}{235}(x - 8)$

$y = 22 + 1.8383x - 14.706$

$y = 7.29 + 1.84x$

Using **4**

Keep at least 5 significant figures here. Round the final answer to 3 s.f.

(b) $y = a + \frac{-18.2}{2956}x$

$4.4 = a + (-0.006\,157\,0)(-7.8)$

$a = 4.3519$

$y = 4.35 - 0.006\,16x$

Using **4**

Now substitute $x = -7.8$ $y = 4.4$. Be careful with the signs. Round the final answer to 3 s.f.

Worked example 2

Andrew (A), Charles (C) and Edward (E) are employed by the Palace Hotel. Each is responsible for one floor of the building and their duties include cleaning the bedrooms. The number of bedrooms occupied on each floor varies from day to day.

The following table shows 10 observations of the number, x, of bedrooms to be cleaned and the time taken, y minutes, to carry out the cleaning. The employee carrying out the cleaning is also indicated.

Employee	A	C	E	E	C	A	A	E	C	C
x	8	22	12	24	19	14	22	16	10	21
y	110	211	132	257	184	165	248	171	97	196

(a) Plot a scatter diagram of the data. Identify the employee by labelling each point.

(b) Calculate the equation of the regression line of y on x. Draw the line on your scatter diagram.

(c) Use your regression equation to estimate the time which would be taken to clean 18 bedrooms.

(d) Calculate the residuals for the three observations when Andrew did the cleaning.

(e) Modify your estimate in part **(c)**, given that the 18 bedrooms are to be cleaned by Andrew.

(a)

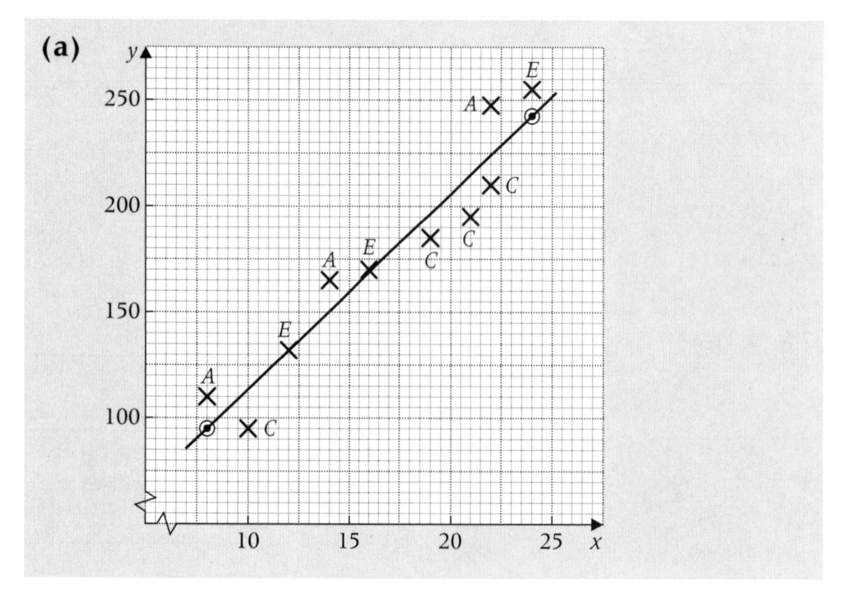

(b) $y = 22.8 + 9.19x$
$x = 8, y = 96.3 \quad x = 24, y = 243.2$

Using **4**

Use your calculator to find the equation. Give your answer to 3 s.f. but retain full accuracy in your calculator for use in later calculations.

Evaluate y at two values of x – one near the lowest and one near the highest observed value.

Plot the points and join with a straight line. (A third point may be evaluated and plotted as a check.)

If the line does not pass through the points on the scatter diagram, there is an error – probably in the equation – possibly in the scatter diagram or the values of y you have evaluated.

(c) $y = 22.77 + 9.186 \times 18 = 188$

(d) $110 - 22.77 - 9.186 \times 8 = 13.7$
$165 - 22.77 - 9.186 \times 14 = 13.6$
$248 - 22.77 - 9.186 \times 22 = 23.1$

(e) All Andrew's residuals are positive (he takes longer than predicted by the equation).
Mean residual for Andrew is
$(13.7 + 13.6 + 23.1)/3 = 16.8$.
Estimated time for Andrew to clean 18 bedrooms
$22.77 + 9.186 \times 18 + 16.8 = 205$ minutes.

7

Worked example 3

A circus company tours the country performing in a marquee which they erect in public parks. At each venue a number of local staff are employed to assist the members of the company in setting up the marquee and other equipment. The manager experimented to see if there is a relationship between the number of local staff employed and the time taken to set up the marquee.

The following data was collected at nine successive venues.

Number of local staff employed, x	2	4	6	7	7	9	9	11	12
Time, minutes, to set up marquee, y	299	282	264	295	258	244	233	218	225

(a) Plot a scatter diagram of these data.

(b) Calculate the equation of the regression line of y on x. Draw the line on your scatter diagram.

(c) If the regression equation is denoted by $y = a + bx$, give an interpretation to each of a and b.

(d) At one of the venues, gale force winds were blowing while the marquee was being erected. Which venue do you think this was? Give a reason for your choice.

(e) At each venue the manager employs as many or more local staff as at the previous venue. Explain why this makes the data less than ideal for assessing the effect of local staff.

(a)

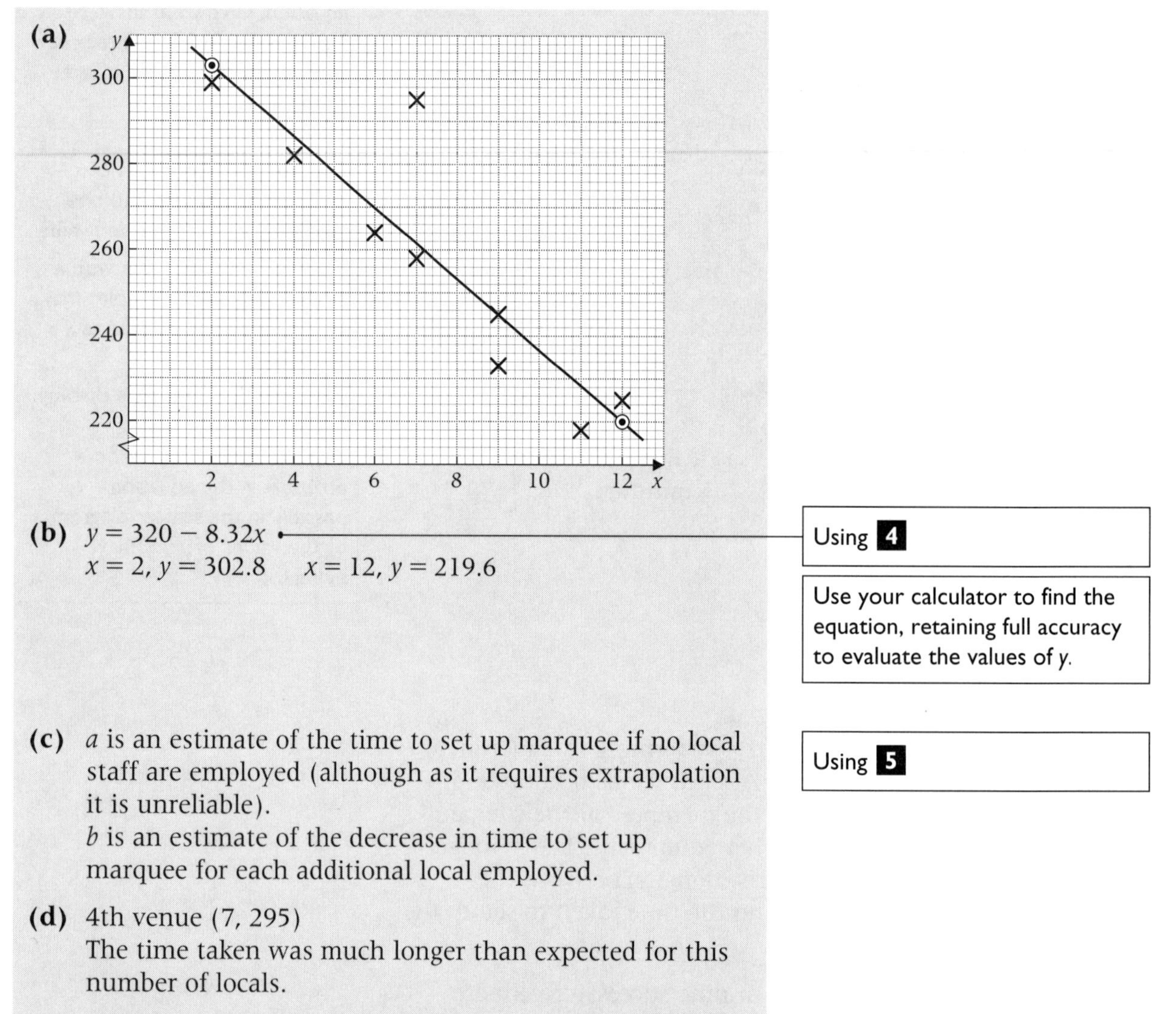

(b) $y = 320 - 8.32x$
$x = 2, y = 302.8 \qquad x = 12, y = 219.6$

Using **4**

Use your calculator to find the equation, retaining full accuracy to evaluate the values of y.

(c) a is an estimate of the time to set up marquee if no local staff are employed (although as it requires extrapolation it is unreliable).
b is an estimate of the decrease in time to set up marquee for each additional local employed.

Using **5**

(d) 4th venue (7, 295)
The time taken was much longer than expected for this number of locals.

(e) As the number of local staff increases over time it is not possible to distinguish between the effect of additional local staff and other possible effects. For example, the company may be becoming more efficient at setting up the marquee.

Worked example 4

A group of students planned to share a house whilst at university. They identified twelve possible houses and used a map to measure the straight line distance, x km, of each house from the Student Union. They then measured the road distance, y km, of each house from the Student Union. The distances are shown below.

House	*A*	*B*	*C*	*D*	*E*	*F*	*G*	*H*	*I*	*J*	*K*	*L*
x	7.5	3.0	23.5	13.4	9.3	9.1	9.8	3.7	17.9	4.7	2.0	2.4
y	8.8	3.2	28.4	16.7	9.5	8.9	12.4	9.9	22.5	4.9	2.5	2.9

(a) Draw a scatter diagram of the data.

(b) A statistician recommended that houses F and H should be omitted from any further analysis as she suspected errors in their data. Discuss why she made this recommendation and, for each house, the strength of the evidence supporting her suspicions.

(c) Calculate the equation of the regression line of road distance on straight line distance after omitting these two houses.

(d) Draw the regression line on your scatter diagram.

(e) Comment on the fact that the regression line does not pass through the origin.

(f) For a geography project, data was collected by travelling a road distance, y km, from the city centre along a main route out of the city and then estimating from a map the straight line distance, x km, from the city centre. Would it be appropriate to calculate the regression line of y on x or of x on y? Explain your answer.

(a)

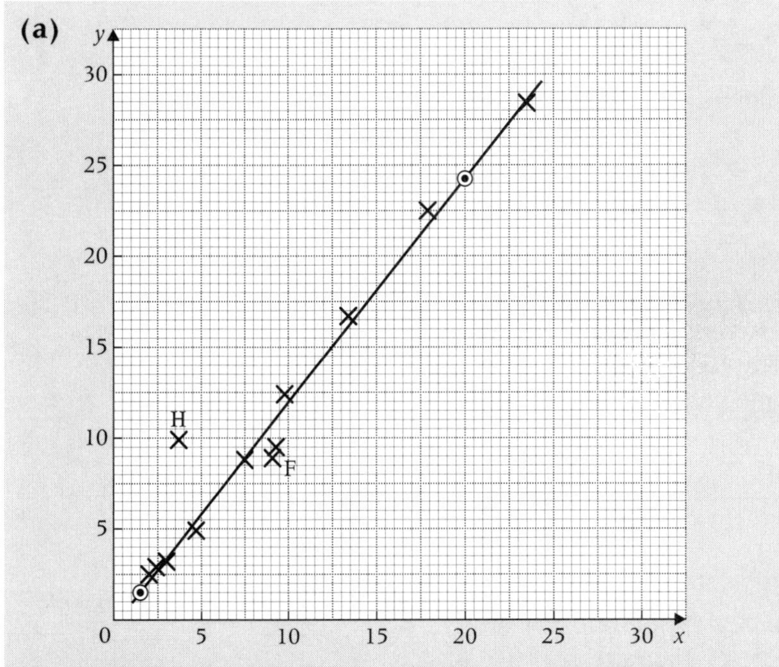

(b) Data for F is definitely incorrect since straight line distance exceeds road distance and this is impossible. Data for H is probably incorrect since road distance is nearly three times straight line distance. This is unlikely but not impossible.

7

(c) $y = -0.447 + 1.24x$

Using **4**

(d) $x = 1.5, y = 1.4 \qquad x = 20, y = 24.4$

Use your calculator to find the equation.

(e) When $x = 0$ (house in Student's Union), $y = 0$.
Regression line passes close to but not through the point.
This could be because:
- relationship is only approximately linear,
- linear relationship does not apply for small values of x,
- calculated value of a depends on particular houses observed and is only an estimate of the population value.

Any one of these reasons would be sufficient for full marks.

(f) Calculate x on y, as in this case the straight line distance x depends on the road distance y.

REVISION EXERCISE 7

1 Calculate the equation of the regression line of y on x in the following cases.

(a)

x	15	4	9	12	17	19
y	13	11	9	10	16	23

(b)

x	15.9	17.2	19.6	21.3	14.0	21.3	32.2	29.1	17.2
y	16.0	38.6	11.2	9.4	18.5	2.8	1.7	9.0	16.4

(c) $S_{xx} = 292$, $S_{xy} = 374$, $\bar{x} = 18$, $\bar{y} = 7$

(d) $S_{xx} = 99.33$, $S_{xy} = -8.325$ and the line passes through the point $(1.22, 2.40)$.

(e) $\Sigma(x-\bar{x})^2 = 211$, $\Sigma(x-\bar{x})(y-\bar{y}) = -300$, $\bar{x} = -3.4$, $\bar{y} = 15.9$

(f) $\Sigma(x-\bar{x})^2 = 0.003\,451$, $\Sigma(x-\bar{x})(y-\bar{y}) = 0.777$ and the line passes through the point $(-2.3, -4.1)$.

2 A local education authority arranges for all children in the area to have their reading ages tested on entering secondary school and again one year later. The following table gives the reading ages, x, on entry and, y, one year later for a sample of nine children attending school A.

x	8.7	11.2	14.4	6.8	12.3	13.1	9.6	10.8	11.2
y	8.8	13.0	16.9	7.2	13.9	15.0	10.4	11.3	13.3

(a) Draw a scatter diagram of this data.

(b) Calculate the equation of the regression line of y on x and draw the line on your scatter diagram.

The same procedure was followed at a neighbouring school B. It was found that at this school the relationship was approximately linear and that the equation of the regression line was $y = 0.7x + 4.3$. The range of values of x was similar to that at school A.

(c) Draw this regression line on your scatter diagram.

The local education authority is particularly keen to improve reading standards.

(d) Which school should it recommend to the parents of a child, about to enter secondary school, with a reading age of:

(i) 7.8 years, (ii) 14.2 years?

Explain your answers.

3 Anne is an athlete who is recovering from injury. Her pulse rate was measured after performing a predetermined number of step-ups in a gymnasium. The measurements were made at weekly intervals. The table below shows the number of step-ups, x, the pulse rate in beats per minute, y, and the week in which the measurement was made.

Week	1	2	3	4	5	6	7	8
x	15	50	35	25	20	30	10	45
y	114	155	132	112	96	105	78	113

(a) Illustrate the data by a scatter diagram. Label the points according to the week in which the measurement was made.

(b) Calculate the equation of the regression line in the form $y = a + bx$. Draw the line on your scatter diagram.

(c) Give an interpretation to the values of a and b.

(d) Use your equation to predict the pulse rate after the athlete performed:

(i) 40 step-ups, (ii) 100 step-ups.

(e) Describe, briefly, the relationship suggested by the scatter diagram between the pulse rate and the number of step-ups.

(f) If 100 step-ups are performed in the 9th week, give two reasons why the pulse rate predicted in part **(d)** is likely to be unreliable.

(g) If 40 step-ups were performed in the 9th week, modify the prediction for the pulse rate made in part **(d)**.

Originally it had been suggested that the pulse rate should be measured after 10, 15 and 20 step-ups on the first visit to the gymnasium, after 25, 30 and 35 step-ups on the second visit, and so on for up to six visits.

(h) Make two criticisms of this suggestion, given that the purpose is to establish the relationship between the number of step-ups and the pulse rate.

4 The following data refer to a particular developed country. The table shows for each year, the annual average temperature, x °C, and the amount, y tonnes, of ice cream produced.

Year	1994	1995	1996	1997	1998	1999	2000	2001	2002	2003	2004
x	9.6	9.3	9.8	10.3	10.1	10.4	10.8	9.7	10.7	9.2	9.8
y	2114	2100	2142	2279	2236	2288	2387	2264	2409	2251	2354

(a) Illustrate the relationship between ice cream production and temperature by a scatter diagram. Label the points according to the year.

(b) Calculate the equation of the regression line of y on x and draw the line on your scatter diagram.

(c) Use your equation to estimate the ice cream production in 2005, given that the average temperature in that year was 10.3 °C.

(d) Calculate the residual for each of the years 2001, 2002, 2003 and 2004. Comment on their values and interpret the pattern shown by the scatter diagram.

(e) Modify your estimate of ice cream production in 2005 in the light of the residuals you have calculated.

Test yourself

What to review

If your answer is incorrect:

1 Calculate the equation of the line of regression of y on x for the following data.

x	2	7	14	9	6	5
y	11	3	−2	−3	4	6

See p 55 Example 1 or review Advancing Maths for AQA S1 p 148

2 A newsagent employs schoolchildren to deliver newspapers. A regression equation is to be calculated connecting p, the number of newspapers to be delivered, and t, the time taken to complete the round. Identify the dependent and the explanatory variable. Justify your choice.

See p 58 Example 4 or review Advancing Maths for AQA S1 p 146

3 The regression equation $y = a + bx$, relates the petrol consumption of a school minibus, y, to the number of pupils carried. Give an interpretation of both a and b.

See p 57 Example 3 or review Advancing Maths for AQA S1 pp 148–149

4 Nasser organises a street collection for a mental health charity. The collection takes place in a large city on a particular Saturday. Volunteers, with collecting tins, stand in busy places and ask passers-by for donations. The following table shows, for ten volunteers, the times, x minutes, they spent collecting together with the amounts, to the nearest pound, y, they collected.

See p 56 Example 2 or review Advancing Maths for AQA S1 pp 147 and 150–153

Test yourself (*continued*)

What to review

Volunteer	*A*	*B*	*C*	*D*	*E*	*F*	*G*	*H*	*I*	*J*
x	65	187	126	52	143	90	157	74	88	195
y	19	53	25	8	30	26	43	19	21	47

(a) Plot a scatter diagram of the data.
(b) Calculate the equation of the regression line of y on x and draw the line on your scatter diagram.
(c) Use your regression equation to estimate the amount collected by a volunteer who collected for:
(i) 100 minutes, (ii) 300 minutes.
(d) Comment on the likely accuracy of each of the estimates you have made in part **(c)**.
(e) Calculate the residuals for volunteers *B* and *C*. Comment on their relative effectiveness as collectors.

Test yourself ANSWERS

1 $y = 11.3 - 1.14x$

2 Explanatory variable is p, dependent variable is t, since time to deliver will depend on the number of papers to be delivered.

3 a is the estimate of petrol consumption if no pupils are carried.
b is the estimate of change in petrol consumption for each additional pupil carried.

4 (a)

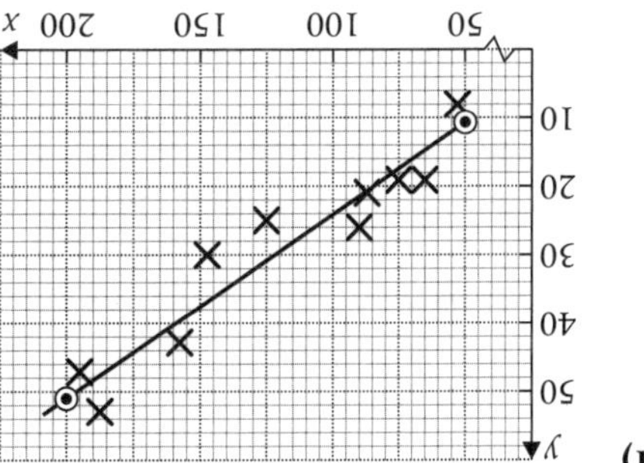

(b) $y = -2.02 + 0.264x$

(c) (i) 24.4 (ii) 77.3

(d) Estimate for 100 minutes likely to be reasonably accurate as it's within observed values and points are quite close to a straight line. Estimate for 300 minutes requires extrapolation and so is unreliable.

(e) *B* 5.58 *C* −6.29
B has positive residual, indicating more collected than expected in given time.
C has negative residual, indicating less collected than expected in given time.
B collected more than twice as much as *C*, partly by collecting for longer and partly by collecting at a faster rate.

7

Examination style paper

Answer **all** questions.
Time allowed: 1 hour 30 minutes

1 The systolic blood pressure, x mm Hg, and the diastolic blood pressure, y mm Hg, of a group of ten patients were measured. The results are shown in the following table.

Patient	1	2	3	4	5	6	7	8	9	10
x	170	129	147	120	136	133	134	158	108	114
y	110	102	104	84	81	95	78	86	87	80

(a) Calculate the product moment correlation coefficient between systolic blood pressure and diastolic blood pressure for these patients. (3 marks)

(b) Interpret your result in part **(a)**. (2 marks)

2 The potencies of a particular brand of sleeping pill may be assumed to be normally distributed with a standard deviation of 0.55. The potencies of a random sample of these pills were as below.

58.3 59.4 60.5 59.2 59.9 60.7 60.3 59.6

(a) Construct a 95% confidence interval for the mean potency of this brand of sleeping pill. (5 marks)

(b) The manufacturer states that the potencies of all sleeping pills of this brand lies in the range 59 to 61.
Comment on this statement in the light of the sample data and your calculations in part **(a)**. (3 marks)

3 A chemist sells two types of bath cube. One type is relaxing and the other type is invigorating. For each cube there is independently a probability of 0.10 that it will be invigorating.

(a) Find the probability that a random sample of 50 cubes will include:
(i) four or fewer invigorating cubes,
(ii) more than one but fewer than 6 invigorating cubes.
(5 marks)

(b) Find the mean and standard deviation of the number of invigorating cubes in a random sample of 50 cubes.
(3 marks)

(c) The owner of a small hotel buys 45 cubes from the chemist. Calculate the probability that these will include exactly 42 relaxing cubes. (4 marks)

The 45 cubes included both relaxing and invigorating cubes.

(d) State, giving a reason, whether or not a binomial distribution will provide an appropriate model for the random variable, R, in each of the following cases;

(i) Seamus, a guest of the hotel, randomly selects one of the 45 cubes. If it is not an invigorating cube, he replaces it and again selects a cube at random. He continues this procedure until he obtains an invigorating cube. The random variable R denotes the number of cubes Seamus selects until he obtains an invigorating cube.

(ii) The owner randomly selects 20 of the 45 cubes and places them in a bowl. The random variable R denotes the number of invigorating cubes the owner places in the bowl. (4 marks)

4 The distance, in kilometres, travelled to work by the employees of a city council may be modelled by a Normal distribution with mean 7.5 and standard deviation 2.5.

(a) Find the probability that the distance travelled to work by a randomly selected employee of the city council is:

(i) less than 11.0 km, (3 marks)

(ii) between 5.5 km and 10.5 km. (4 marks)

(b) Find d such that 10% of the council's employees travel less than d kilometres to work. (4 marks)

(c) Find the probability that the mean distance travelled to work by a random sample of 6 of the council's employees is less than 5.0 km. (4 marks)

5 A meeting of a statistical society was attended by 95 people. The following table summarises their mode of transport to the meeting and their gender.

	Mode of Transport			
	Bicycle	**Public Transport**	**Private car**	**Walk**
Female	3	26	9	2
Male	4	32	16	3

(a) One person is selected at random to record the decisions of the meeting.
B denotes the event that the person selected came to the meeting by bicycle.
F denotes the event that the person selected is female.
B' denotes the event not B.

Determine:

(i) P(B), (1 mark)

(ii) P($B \cap F$), (1 mark)

(iii) P($B | F$), (2 marks)

(iv) P($B' \cup F$). (2 marks)

(b) Define an event which is mutually exclusive to B. (2 marks)

(c) Two people were selected at random (without replacement) to make tea. Determine the probability that:

(i) they are both males who came to the meeting by public transport, (2 marks)

(ii) exactly one is a male who came to the meeting by public transport. (3 marks)

6 Carina obtains cash from an ATM (cash machine). She records, for ten successive visits the amount, £x, withdrawn, and the number of hours, y, until her next visit to an ATM.

Withdrawal	1	2	3	4	5	6	7	8	9	10
x	40	10	100	110	120	150	20	90	80	130
y	56	62	195	330	94	270	48	196	214	286

(a) Draw a scatter diagram of the data. (3 marks)

(b) Calculate the equation of the regression line of y on x and draw it on your scatter diagram. (6 marks)

(c) Calculate the residuals for withdrawals 4 and 5. (3 marks)

(d) (i) Carina made one withdrawal immediately before going on a weekend visit to Edinburgh. Identify the most likely withdrawal, giving a reason. (2 marks)

(ii) Following another withdrawal, Carina was confined to bed for several days with a heavy cold. Identify the most likely withdrawal, giving a reason. (2 marks)

(e) Interpret, in context, the gradient of the regression line. (2 marks)

Answers

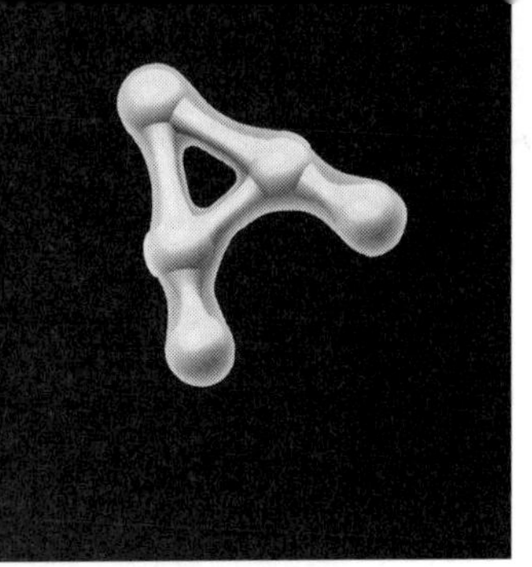

Revision exercise 1

1 Mode 3, median 3, mean 3.2

2 **(a)** Range 9, interquartile range 2.5, standard deviation 2.21 (2.16 accepted)

(b) (i) Range. Highest value decreases by 2 while lowest stays the same.
(ii) Interquartile range. The only change is to the highest value; quartiles are unchanged.

3 4.25, 1.61 (1.59 accepted)

4 **(a)** Mean 7.1 years, standard deviation 4.3 years

(b) Mean 7.6 years, standard deviation 4.3 years

5 **(a)** Mean 9.7 °C, standard deviation 1.4 °C (1.3 accepted)

(b) Mean 49.4 °C, standard deviation 2.4 °C

6 **(a)** Variety A: Mean 86.7 g, standard deviation 5.73 g (5.65 accepted)
Variety B: Mean 88.7 g, standard deviation 11.2 g (11.1 accepted)

(b) Variety B apples are about 2 g heavier on average than variety A. There is much more variation in weights for B while most variety A apples weigh between 80 and 95 grams.

Revision exercise 2

1 **(a)** $\frac{1}{3}$ **(b)** $\frac{1}{4}$ **(c)** $\frac{1}{52}$

2 **(a)** 0.67 **(b)** 0.33

3 **(a)** 0.064 **(b)** 0.784

4 **(a)** $\frac{1}{16}$ **(b)** $\frac{7}{16}$ **(c)** $\frac{1}{4}$ **(d)** $\frac{3}{16}$

5 **(a)** 0.729 **(b)** 0.05 **(c)** 0.58

6 **(a)** (i) 0.55 (ii) 0.45

(b) 0.425

7 **(a)** $\frac{1}{9}$ **(b)** $\frac{5}{6}$ **(c)** $\frac{5}{7}$ **(d)** $\frac{1}{5}$

Revision exercise 3

1 **(a)** 0.650 **(b)** 0.383 **(c)** 0.150 **(d)** 0.700 **(e)** 0.650

(f) Mean 3, standard deviation 1.45

2 **(a)** (i) 0.842 (ii) 0.351 **(b)** 0.153

3 **(a)** (i) 0.0874 (ii) 0.218 (iii) 0.681 **(b)** (i) 0.314 (ii) 0.918 (iii) 0.0173

4 **(a)** (i) 0.0875 (ii) 0.161 **(b)** Mean 8, standard deviation 2.53

5 **(a)** (i) 0.552 (ii) 0.296 (iii) 0.194

(b) Not binomial, as n (number of sandwiches chosen) is not fixed.

6 **(a)** (i) 0.736 (ii) 0.292 (iii) 0.908

(b) (i) 0.4275 (ii) 0.393 (iii) 0.154

7 **(a)** (i) 0.425 (ii) 0.102 (iii) 0.369 (iv) 0.204

(b) £8

(c) Some people more likely to buy than others (p not constant)/people may be influenced by whether or not others buy (not independent).

8 **(a)** (i) 0.420 (ii) 0.745

(b) 0.0426

(c) Mean 2.8, standard deviation 1.30

(d) (i) Mean 2.8, standard deviation 2.27 (2.24 accepted)
(ii) Although the means are the same the standard deviations are very different, suggesting the distribution is not binomial and so Aaron's belief is not supported.

Revision exercise 4

[Some answers may vary slightly if rounding is used in order for probabilities to be obtained from Table 3. Where appropriate 4 s.f. have been given.]

1 **(a)** (i) 0.998 (ii) 0.234 (iii) 0.949 (iv) 0.572

(b) (i) 175.6 cm (ii) 164.0 cm

2 **(a)** 0.918 **(b)** 0.0131 **(c)** 0.733 **(d)** 576.6 g

3 **(a)** (i) 29.0 min (ii) 15.1 min (iii) 0.0668 **(b)** 28.7 min

4 **(a)** (i) 0.841 (ii) 0.369 (iii) 0.0626 **(b)** 12.2 min

5 **(a)** (i) 0.369 (ii) 0.0912 (iii) 0.247

(b) 0.982 **(c)** $\mu = 233.2$ g, $\sigma = 14.1$ g

6 **(a)** (i) 0.747 (ii) 0.495 **(b)** 0.932 **(c)** 0.159 **(d)** 240.1 ml

7 **(a)** 0.894 **(b)** 08.11 am (or 08.10 am)

8 **(a)** (i) 0.280 (ii) 0.202 **(b)** (22.3, 61.7) min **(c)** 0.734

Revision exercise 5

[It is not always appropriate to give the limits of confidence intervals to 3 s.f. Answers have generally been given to one more significant figure than the raw data. Units have been omitted from the numerical answers, but included where interpretation is required.]

1 **(a)** 39.44–49.13 **(b)** 38.07–50.50 **(c)** 36.88–51.69 **(d)** 34.55–54.02 (e) 32.61–55.97

2 102.70–112.30

3 **(a)** 98.07–98.53

(b) The upper limit of the confidence interval is well below 100 grams. Hence there is strong evidence that the mean weight is less than 100 grams, and that the consignment is underweight (although some individual jars may weigh more than 100 grams).

4 **(a)** 294.7–520.5 **(b)** 0.1

5 **(a)** 663.0–677.0

(b) The central limit theorem states that the mean of a large sample from any distribution will be approximately normally distributed. This means that the confidence interval in part **(a)** can be calculated even though it is not known whether the fat contents follow a Normal distribution.

(c) 601.4–738.6

(d) The confidence interval provides evidence that the mean fat content is less than 680 g but many individual packages will contain more than 680 g of fat.

6 **(a)** $\bar{x} = 9.21, s = 3.2100$ **(b)** 8.29–10.13 **(c)** 1.85 **(d)** (i) 96.3 (ii) 83.6

7 **(a)** $\bar{x} = 719, s = 240$

(b) 681.5–756.7

(c) The confidence interval provides evidence that the mean time spent watching television by 12-year-old children exceeds 10 hours (600 minutes).

(d) (i) No change, as sample is large the central limit theorem states that the sample mean will be approximately normally distributed whatever the parent distribution.
(ii) The conclusion would be unreliable as the sample could be completely unrepresentative – for example all 12-year-olds in sample may be from cities and may have different viewing habits from 12-year-olds in small towns and rural areas.

Revision exercise 6

1 **(a)** 0.843

(b) −0.612

(c) 0.566

(d) −0.846

(e) −0.675

(f) 0.715

2 **(a)** −0.95 (−0.75 to −0.99)

(b) 0.8 (0.5 to 0.95)

(c) −0.95 (−0.75 to 0.98)

(d) 0.0 (−0.2 to 0.2)

3 **(a)** (i) 0.929
(ii) Value suggests that towns with a large number of solicitors also tend to have had a large number of cars stolen last week.

(b) No evidence that most car thieves are solicitors. In this case the most likely explanation of the association is that the larger the town the more solicitors and the more cars stolen.

4 **(a)** 0.209

(b) The value is small, showing little evidence of linear association between the estimates of Henri and Michelle.

5 **Statement 1** – probably incorrect – towns with high percentage with qualifications will probably tend to have a low percentage unemployed. This will result in a negative correlation coefficient.
Statement 2 – definitely incorrect – correlation coefficient cannot be greater than one.
Statement 3 – plausible – the larger the town the more mathematics teachers and the more crimes of violence.

6 **(a)**

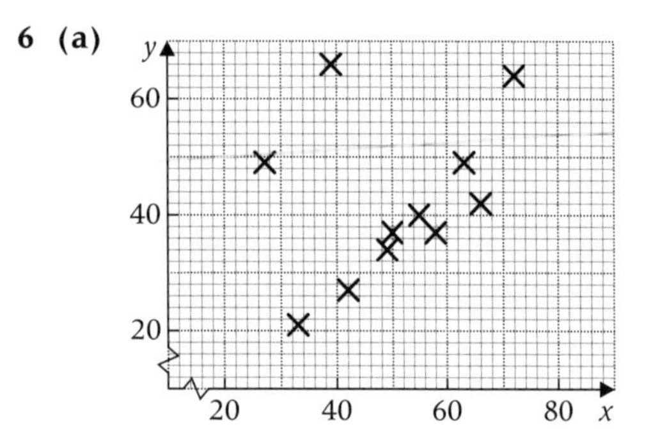

(b) 0.296, little evidence of linear association between left and right hand times.

(c) *E* and *G*, both outliers on scatter diagram, 0.926.

(d) Correlation coefficient suggests strong linear association between left and right hand times among right handed operators.

7 **(a)**

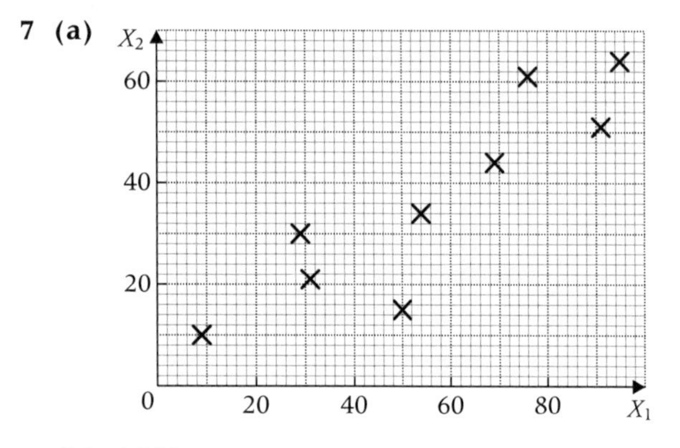

(b) 0.891

(c) There appears to be a strong linear association between X_1 and X_2 with high values of X_1 associated with high values of X_2. The relationship between X_3 and X_4 appears to be almost exactly linear with just one employee having values a little way from the line. High values of X_3 are associated with high values of X_4. No evidence of a relationship between X_5 and X_6. The high value of the correlation coefficient being due to one outlier.

Revision exercise 7

1 **(a)** $y = 4.55 + 0.720x$

(b) $y = 36.5 - 1.09x$

(c) $y = -16.1 + 1.28x$

(d) $y = 2.50 - 0.0838x$

(e) $y = 11.1 - 1.42x$

(f) $y = 514 + 225x$

2 (a) (c)

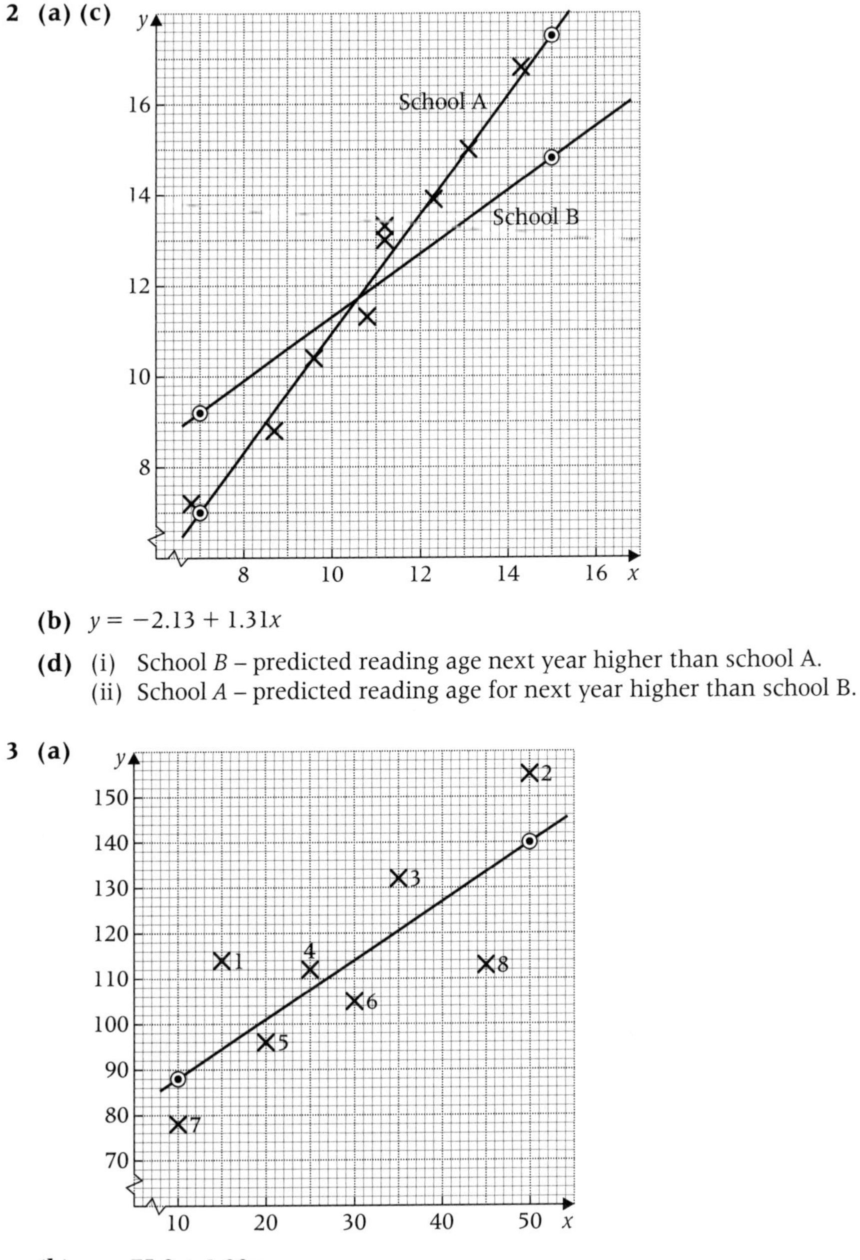

(b) $y = -2.13 + 1.31x$

(d) (i) School *B* – predicted reading age next year higher than school A.
(ii) School *A* – predicted reading age for next year higher than school B.

3 (a)

(b) $y = 75.9 + 1.29x$

(c) a is an estimate of pulse rate before any step-ups have been done.
b is an estimate of increase in pulse rate for each additional step-up.

(d) (i) 128 (ii) 205

(e) Pulse rate increases as number of step-ups increases. For a given number of step-ups pulse rate is greater in early weeks than in later weeks (as athlete becomes fitter).

(f) Estimate involves extrapolation. Pulse rate likely to be lower because athlete is now fitter.

(g) 108 (100–110 acceptable answers)

(h) Doing more than one test on each visit to gym may affect the results (e.g. pulse rate after 35 step-ups may be affected by the fact that 25 and 30 step-ups have been undertaken on the same visit).
The effects on pulse rate of increasing the number of step-ups will be confounded (inextricably mixed up) with the effect of the athlete's increasing fitness.

4 (a)

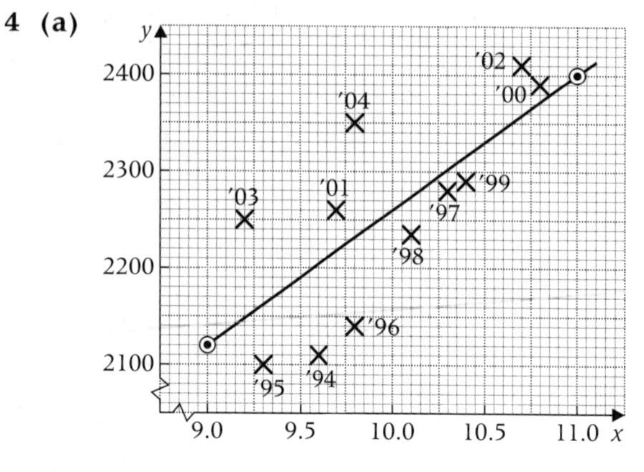

(b) $y = 851 + 141x$ **(c)** 2300 tonnes

(d) 45.7, 49.7, 103.2, 121.6
All residuals are positive and appear to be increasing. Scatter diagram suggests that ice cream consumption increases as the average temperature increases, and there is also a steady increase over time.

(e) 2440 (2420–2470)

Practise paper

1 (a) 0.552

(b) Moderate correlation. Some evidence that high systolic blood pressure is associated with high diastolic blood pressure.

2 (a) 59.36–60.12

(b) Confidence interval suggests that the mean potency is in the range 59 to 61, but the sample shows that the potency of some individual tablets lies outside this range.

3 (a) (i) 0.431 (ii) 0.582 **(b)** mean 5, standard deviation 2.12 **(c)** 0.170

(d) (i) Not binomial, as the number of trials is not constant. (ii) Not binomial, as p is not constant.

4 (a) (i) 0.919 (ii) 0.673 **(b)** 4.30 km **(c)** 0.0072

5 (a) (i) 0.0737 (ii) 0.0316 (iii) 0.075 (iv) 0.958

(b) Various answers, e.g. the person selected walked to the meeting **(c)** (i) 0.111 (ii) 0.452

6 (a)

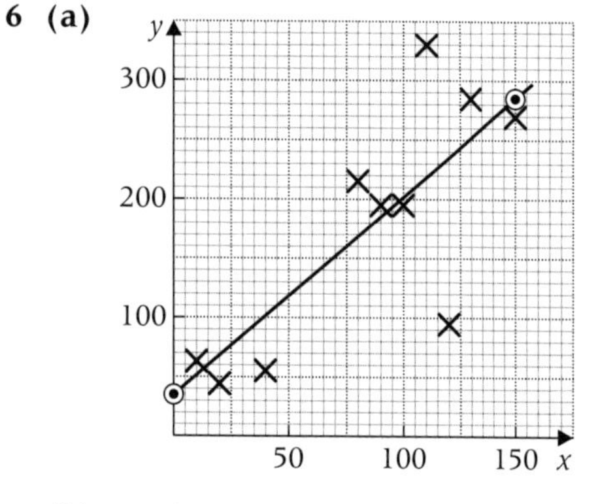

(b) $y = 30.3 + 1.70x$ **(c)** 112, −141

(d) (i) withdrawal 5 – spent a lot of cash in a relatively short time
(ii) withdrawal 4 – spent a small amount of cash in a relatively long time

(e) Estimate of the number of hours each £ withdrawn lasts.